PROJECT AIR FORCE

T0302653

A New Methodology for Conducting Product Support Business Case Analysis (BCA)

With Illustrations from the F-22 Product Support BCA

Frank Camm, John Matsumura, Lauren A. Mayer, Kyle Siler-Evans

Prepared for the United States Air Force

For more information on this publication, visit www.rand.org/t/RR1664

Library of Congress Cataloging-in-Publication Data is available for this publication.
ISBN: 978-0-8330-9633-3

Published by the RAND Corporation, Santa Monica, Calif.
© Copyright 2017 RAND Corporation
RAND® is a registered trademark.

Support RAND
Make a tax-deductible charitable contribution at
www.rand.org/giving/contribute

www.rand.org

Preface

The F-22 program was postured to rely heavily on contractor logistics support during the acquisition phase of its life cycle. In 2009, the F-22 System Program Office led a business case analysis (BCA) of F-22 sustainment. Based on the findings of that BCA, in 2010, the secretary of the Air Force decided to transition most functions, other than supply chain management, from the sustainment contractors to the government. The secretary directed that the feasibility of transitioning supply chain management be re-examined in three to five years. (Section 805 of the fiscal year [FY] 2010 National Defense Authorization Act requires revalidation of BCAs of product support strategies for major weapon systems be revalidated at least every five years.)

In 2014, the F-22 System Program Office began a second product support BCA for the F-22 air vehicle and F119 engine and asked RAND Project AIR FORCE (PAF) to support the BCA. The BCA has included, among other things, assessment of U.S. Air Force progress in implementing recommendations from the 2010 Product Support BCA, identification of additional F-22 sustainment elements that could be transitioned to organic support in 2018 and beyond, and assessment of a variety of alternate support strategies (including new applications of performance-based logistics agreements).

In the course of supporting that BCA, PAF developed a new approach to assessing and comparing the courses of action (COAs) that a BCA uses to define policy alternatives. Although this approach does not represent the way the Air Force typically conducts BCAs, this method is compliant with Office of Management and Budget and U.S. Department of Defense policy and it offers a new way to identify and assess sources of risk that can delay or even prevent full implementation of a COA. The approach integrates this assessment of risk with cost analysis in a way that allows the user to characterize each COA and each COA element in terms of dollars of net present value (NPV) in the same way that the government routinely assesses the NPV of investment alternatives. The approach also examines many potential states to capture the full risk effects of competing COAs. The resulting dollar-based figure of merit makes it easier for senior decisionmakers to compare COAs and to consider COA adjustments as they move toward decisions about product support.

This document draws on the F-22 product support BCA for examples and illustrations of the methods documented here, but it is primarily meant to inform personnel associated with future COAs. It should interest all those involved in conducting, overseeing, and reacting to product support BCAs. It should also interest analysts responsible for conducting a broader class of cost benefit analysis, in which risk assessment is integral to analysis and analysts seek to compare the performance of investment options across many different potential states.

The research reported here was sponsored by Maj Gen Dwyer L. Dennis, executive officer of the Air Force Life Cycle Management Center, Fighter-Bomber Directorate, and conducted within the Resource Management Program of PAF as part of an FY 2014 project, "Support for FY 2015 F-22 Sustainment Business Case Analysis." The new BCA methodology proposed in this report is not endorsed or approved by the U.S. Air Force or the U.S. Government.

RAND Project AIR FORCE

RAND Project AIR FORCE (PAF), a division of the RAND Corporation, is the U.S. Air Force's federally funded research and development center for studies and analyses. PAF provides the Air Force with independent analyses of policy alternatives affecting the development, employment, combat readiness, and support of current and future air, space, and cyber forces. Research is conducted in four programs: Force Modernization and Employment; Manpower, Personnel, and Training; Resource Management; and Strategy and Doctrine. The research reported here was prepared under contract FA7014-06-C-0001.

Additional information about PAF is available on our website: http://www.rand.org/paf/

This report documents work originally shared with the U.S. Air Force on November 30, 2015. The draft report, issued in March 2016, was reviewed by formal peer reviewers and U.S. Air Force subject-matter experts.

Contents

Figures

Tables

Summary

U.S. Department of Defense (DoD) guidance states that a product support business case analysis (BCA) "aids decision making by identifying and comparing alternatives." It does this by "examining the mission and business impacts (both financial and nonfinancial), risks, and sensitivities" relevant to choosing among alternative courses of action (COAs) for supporting a product. "Other names for a BCA are Economic Analysis, Cost-Benefit Analysis, and Benefit-Cost Analysis. Broadly speaking, a BCA is any documented, objective, value analysis exploring costs, benefits, and risks."[1]

DoD guidance directs that any BCA consider the costs, benefits, and risks associated with each COA under review. Costs relevant to each COA include nonrecurring costs, which mainly occur in the opening years of the time period considered in the analysis, and recurring costs, which occur over the whole period analyzed. Benefits define what effect on performance the government gets for the money it spends for each COA. DoD guidance highlights, among many other possibilities, availability, reliability, supportability, manageability, versatility, and system life. The guidance defines *risk* as an "undesirable implication of uncertainty" that "can be a factor in eliminating or reducing the value of an alternative that is otherwise highly evaluated."[2]

This report describes a new analytic approach to a BCA that is based on traditional benefit-cost analysis and consistent with Office of Management and Budget (OMB) and DoD guidance on conducting benefit-cost analysis.[3] The approach described here views each COA in a product support BCA as a distinct project with an associated sequence of annual flows of benefits (think "nonfinancial mission and business impacts") and costs (think "financial mission and business impacts"); examines how this sequence might differ under alternative assumptions (think "sensitivities") and in alternative futures; monetizes the flows in this sequence; and calculates information on the net present value (NPV) of each COA that decisionmakers can use to

---

[1] Office of the Principal Deputy Assistant Secretary of Defense for Logistics and Materiel Readiness, "Introduction," Section 1.1, *DoD Product Support Business Case Analysis Guidebook*, Washington, D.C.: Office of the Secretary of Defense, 2011.

[2] Office of the Principal Deputy Assistant Secretary of Defense for Logistics and Materiel Readiness, "Risk Analysis in a BCA," Section 4.7.1.1, *DoD Product Support Business Case Analysis Guidebook*, Washington, D.C.: Office of the Secretary of Defense, 2011, and Office of the Principal Deputy Assistant Secretary of Defense for Logistics and Materiel Readiness, "Risk Classification," Section 4.7.1.2, *DoD Product Support Business Case Analysis Guidebook*, Washington, D.C.: Office of the Secretary of Defense, 2011.

[3] OMB, *Guidelines and Discount Rates for Benefit-Cost Analysis of Federal Programs*, Circular No. A-94 (revised), Washington, D.C., 1992, and Director, Cost Assessment and Program Evaluation, "Economic Analysis for Decision-Making," DoD Instruction 7041.03, Washington, D.C.: Department of Defense, September 9, 2015.

compare alternative COAs. Commercial companies routinely use this standard approach to compare business alternatives.[4]

What Is New About the Approach Described Here

We designed this approach to be as helpful as possible to the senior government leaders who will receive the findings of a BCA and act on them, departing from the approach we have seen in past product support BCAs. Table S.1 summarizes important differences between the approach described here and the approach we have often seen used in the past. We believe that this new approach offers three significant improvements on the traditional approach.

First, it uses dollar measures of value rather than measures of value based on abstract "scoring and weighting." This allows us to avoid many of the problems that DoD BCA guidance associates with scores and weights, including explaining (1) the practical meaning of abstract scoring scales, (2) the differences in the policy relevance of different factors scored, and (3) the normalization used to make systems of scores and weights compatible.[5] Senior leaders routinely use dollar values to inform choices among alternatives, especially in the context of the planning, programming, and budgeting processes.

Second, it offers a natural way to integrate thinking about risks with thinking about benefits and costs and uses standard project evaluation tools used throughout the government and private industry.

Third, it also offers a natural way to integrate COA implementation challenges with information about COA benefits and costs. We have found repeatedly that senior leaders have limited interest in a COA that the government will likely have difficulty implementing. The approach described here informs them directly about how implementation difficulties affect the value of a COA.

---

[4] See, for example, Harry F. Campbell and Richard P. C. Brown, *Benefit-Cost Analysis: Financial and Economic Appraisal Using Spreadsheets*, Cambridge, UK: Cambridge University Press, 2003; Matthew D. Adler and Eric A. Posner, eds., *Cost-Benefit Analysis: Economic, Philosophical, and Legal Perspectives*, Chicago, Ill.: University of Chicago Press, 2001.

[5] These challenges appear in the following sections of Office of the Principal Deputy Assistant Secretary of Defense for Logistics and Materiel Readiness, 2011: 4.4.3.1, "Quantitative and Qualitative Values"; 4.4.3.2, "Scoring and Weighting"; 4.4.3.3, "Quantifying Qualitative Values"; 4.4.3.4, "Normalization"; and 4.4.3.5, "Rank Ordering/Prioritization."

Table S.1. How the RAND Approach Differs from That Typically Used in Support BCAs

RAND Approach	Traditional Approach
Provide a fully integrated assessment of the costs, benefits, and risks relevant to each COA. Use the standard cost-benefit guidance of OMB Circular A-94 to treat each COA as a formal investment alternative.[a]	Assess costs, benefits, and risks separately. State cost in dollar terms. Summarize the probability and impact associated with each individual source of risk. Report subjective inputs in a summary five-by-five matrix that associates each source of risk with one of five levels of probability and one of five levels of impact.
Formally recognize the pervasive presence of uncertainty about the future. Capture this by presenting a subjective probability distribution for NPV for each COA.	Develop point estimates of cost and benefit. Rely on assessments of the probability and impact associated with each individual source of risk to convey implications of uncertainty.
Use sensitivity analysis to explore idiosyncratic uncertainties not likely to be captured in the subjective probability distribution for each COA.	Use sensitivity analysis to explore discrete uncertainties relative to some base case.
Focus on ensuring that every COA achieves a threshold target associated with the primary benefit highlighted in the BCA ground rules. Monetize the cost of ensuring that each COA achieves the threshold. Provide a framework to inform decisionmakers of the monetary cost of pursuing secondary benefits by preferring any COA other than the one that offers the highest NPV while achieving the primary threshold target.	Identify several—potentially many—benefits. Elicit information on their relative importance to decisionmakers. Score each COA on each benefit using scales normalized to be compatible with the measures of relative importance used. Identify the COA with the highest weighted score. Do not consider risks when calculating this score.
Use formal risk assessment methods to elicit any subjective inputs in a way that minimizes the opportunity for introducing bias. Use formal, transparent, repeatable methods to translate these inputs into quantitative figures of merit.	Guidance focuses on seeking inputs that properly reflect the scores and scales used to calculate a weighted score for each COA. It does not address methods that could unintentionally bias these inputs.

[a] OMB, 1992.

A Step-by-Step Summary of the Approach

Twelve steps summarize the approach presented in this document.

1. Identify the primary attribute of performance that senior decisionmakers care most about as they compare COAs in a BCA. In a product support BCA, this attribute is likely to be the material status of the products relevant to warfighters—for example, the operational availability of weapon systems.

2. Clarify which COA elements decisionmakers want most to compare. *COA elements* are discrete proposals within a COA that affect distinct activities and that each present potentially different sets of issues of performance, cost, and risk. Within a COA, examples of COA elements might include insourcing material management for selected items, transferring material management of a particular component to an existing cross-system performance-based logistics agreement, or improving the performance of a contractor-controlled process. Identify data available to support comparisons. Structure the BCA around COA elements that are important to policymakers and have enough data for detailed analysis.

3. Focus risk assessment on factors that can prevent the government from realizing the full net benefits from pursuing any COA element. Treat risk as uncertainty about how far any COA element might fall short of achieving its full net benefit.

4. Define a baseline COA over the period of analysis. This COA describes the future if the government maintained pre-BCA policies, processes, and practices and made no changes—the COA that the government can "choose" by doing nothing.

5. Identify the real (adjusted for any future inflation) savings that each COA element can generate *relative* to the baseline COA if successfully implemented to ensure the target level of performance that policymakers set as the most important attribute of performance. Savings might come, for example, from substituting lower-cost government personnel for contractor personnel, improving management of second-tier vendors, or improving contractor processes. Express these costs and savings in terms of annual cost flows over the period of analysis.

6. Identify other, secondary attributes of performance important to decisionmakers. Examples might include enhanced ability to introduce new technologies, enhanced agility of support for deployed forces, or insight into system support. Assess the level of performance each COA achieves for each of these secondary attributes of performance *relative* to the baseline COA.

7. Identify key sources of risk or risk drivers (e.g., lack of access to relevant technical data or personnel, inability to execute a performance-based logistics agreement) and methods to assess their effects on COA element implementation. Use initial, open-ended discussions to identify risk drivers relevant to each COA element and the channels of influence through which each driver affects COA element implementation. If the risk drivers and channels of influence are similar to those in previous BCAs, use those BCAs to identify formal methods for modeling how risk drivers affect implementation. If not, construct tailored expert models of the risk drivers and their channels of influence. Use these models to determine what parameters must be valued to compare COA elements and COAs. Examples might include the relative importance of a risk driver or the length of time needed to mitigate associated problems. Use the models to determine the structure of an aggregation function that can translate information on individual risk drivers into information on the potential implementation of each COA element.

8. Collect available, objective, quantitative information on the values of parameters identified in Step 7. Where that is not available, use formal risk assessment methods to collect information from subject-matter experts. These methods should govern precisely what information analysts collect and what methods they use to collect it. This information should include at least the following: (a) which risk drivers are relevant to each COA element, (b) whether the problems associated with any risk driver must be fully mitigated to ensure successful COA element implementation, (c) the relative importance of other risk drivers to successful implementation, (d) the probability that the government can mitigate the problems associated with each risk driver, (e) the real monetary cost of mitigating problems associated with each risk driver, and (f) the appropriate risk-free real discount factor to use when computing NPVs for COA elements in individual futures.

9. Use Monte Carlo simulation to construct multiple futures for each COA element.[6] The Monte Carlo analysis first uses random draws based on the probability of successfully

---

[6] Monte Carlo simulation allows an analyst to generate random scenarios that are consistent with the analyst's key assumptions about risk. It "quantitatively describe[s] the uncertainty surrounding . . . key project variables as probability distributions, and . . . calculate[s] in a consistent manner its possible impact on the project's value" (Savvakis Savvides, "Risk Analysis in Investment Approach," *Project Appraisal*, Vol. 9, No. 1, March 1994, p. 3).

mitigating problems associated with all risk drivers relevant to a COA element to determine the degree of success realized in one future. When draws are complete for all risk drivers in one future, the analysis calculates the NPV achieved in this future. The Monte Carlo analysis repeats such sequences of draws to construct many futures and calculate the NPV achieved in each one. It finally uses the NPVs identified in these futures to construct a subjective probability distribution of NPV for each COA element.

10. Use Monte Carlo simulation to construct multiple futures for each COA. This analysis first treats the outcomes for the elements within a COA as statistically independent of one another and uses random draws based on the subjective probability distributions constructed in Step 9 to determine the NPV of the COA in one future. When draws are complete for all elements of a COA in one future, the analysis sums the NPVs for these elements to calculate the NPV for that future. The Monte Carlo analysis repeats such sequences of draws to construct many futures and calculate the NPV of each COA achieved in each future. It finally uses the NPVs identified in these futures to construct a subjective probability distribution of NPV for each COA.

11. Calculate summary statistics that convey information from the subjective probability distributions of NPV constructed in Steps 9 and 10 that interest decisionmakers. Three statistics of interest include (a) the expected value for decisionmakers who are risk neutral, (b) the probability of a negative NPV for decisionmakers who are loss averse, and (c) a subjective confidence interval for decisionmakers who are risk averse or who want a sense of the level of uncertainty associated with the NPV for any COA or COA element.

12. Assess the relevance of secondary performance attributes identified in Step 6. Determine whether competitive COAs achieve significantly different levels of performance as measured in terms of these secondary attributes.[7] If they do, determine which COA the information generated in Steps 9 through 11 favors and ask whether information on any of these secondary attributes favors a different COA. If not, recommend the COA preferred on the basis of information in Steps 9 through 11. If a secondary attribute points to a different choice, compare the distributions of NPV for the COA preferred in Steps 9 through 11 and the distribution preferred based on the secondary attribute. To do this, define a new variable: the difference in any future between the NPVs for these two COAs. Use Monte Carlo analysis to construct a subjective probability distribution for this difference. To do this, use random draws based on the subjective probability distributions constructed in Step 10 to calculate the values of this difference in many futures. Construct a subjective probability distribution for this difference. This distribution identifies how much NPV decisionmakers would have to be willing to forego to choose the COA favored by the secondary performance attribute (from Step 6) rather than the one favored by the primary attribute (from Step 1). If desired, calculate the summary statistics described in Step 11 for this new distribution.

Savvides provides practical guidance on how to implement the approach first described in David B. Hertz, "Risk Analysis in Capital Investment," *Harvard Business Review*, Vol. 42, No. 1, February 1964, p. 95.

[7] Analysts can use an analogous approach to examine the relevance of alternative performance attributes to competitive COA elements.

Illustrative Examples from the 2015 F-22 Product Support BCA

We developed this approach in the context of the second F-22 product support BCA and use concrete examples from that BCA throughout to illustrate its elements. The approach described here is one that can be applied in other product support BCAs in the future. We direct those seeking details on the F-22 BCA to documents that describe that study in detail.[8] This document emphasizes a general methodology more than findings specific to F-22 sustainment.

---

[8] Details on the F-22 Product Support BCA are not available to the general public. Please contact Michael Boito (boito@rand.org) or Kristin Lynch (lynch@rand.org) at the RAND Corporation for information on what materials can be shared.

1. Introduction

U.S. Department of Defense (DoD) guidance states that a product support business case analysis (BCA) "aids decisionmaking by identifying and comparing alternatives." It does this by "examining the mission and business impacts (both financial and nonfinancial), risks, and sensitivities" relevant to choosing among alternative courses of action (COAs) for supporting a product. "Other names for a BCA are Economic Analysis, Cost-Benefit Analysis, and Benefit-Cost Analysis. Broadly speaking, a BCA is any documented, objective, value analysis exploring costs, benefits, and risks."[1]

DoD guidance directs that any BCA consider the costs, benefits, and risks associated with each COA under review. Costs relevant to each COA include nonrecurring costs, which mainly occur in the opening years of the time period considered in the analysis, and recurring costs that occur over the whole period analyzed. Benefits define what effect on performance the government gets for the money it spends for each COA. DoD guidance highlights—among many other possibilities—availability, reliability, supportability, manageability, versatility, and system life. The guidance defines risk as an "undesirable implication of uncertainty" that "can be a factor in eliminating or reducing the value of an alternative that is otherwise highly evaluated."[2]

DoD guidance recommends using a "scoring and weighting methodology, such as Value Focus [sic] Thinking and Analytical Hierarchy Process."[3] This document describes an alternative, economic analytic approach that is based on traditional benefit-cost analysis and consistent with Office of Management and Budget (OMB) and DoD guidance on how to conduct benefit-cost analysis that can be used in BCAs to help senior leaders make important strategic resourcing decisions.[4]

---

[1] Office of the Principal Deputy Assistant Secretary of Defense for Logistics and Materiel Readiness, "Introduction," Section 1.1, *DoD Product Support Business Case Analysis Guidebook*, Washington, D.C.: Office of the Secretary of Defense, 2011.

[2] Office of the Principal Deputy Assistant Secretary of Defense for Logistics and Materiel Readiness, "Risk Analysis in a BCA," Section 4.7.1.1, *DoD Product Support Business Case Analysis Guidebook*, Washington, D.C.: Office of the Secretary of Defense, 2011, and Office of the Principal Deputy Assistant Secretary of Defense for Logistics and Materiel Readiness, "Risk Classification," Section 4.7.1.2, *DoD Product Support Business Case Analysis Guidebook*, Washington, D.C.: Office of the Secretary of Defense, 2011.

[3] Office of the Principal Deputy Assistant Secretary of Defense for Logistics and Materiel Readiness, "Evaluation Criteria," Section 4.4.3, *DoD Product Support Business Case Analysis Guidebook*, Washington, D.C.: Office of the Secretary of Defense, 2011. For a description of value-focused thinking, see Ralph L. Keeney, *Value-Focused Thinking*, Cambridge, Mass.: Harvard University Press, 1996. For a description of the analytic hierarchy process, see Thomas L. Saaty, *The Analytic Hierarchy Process: Planning, Priority Setting, Resource Allocation*, New York: McGraw Hill, 1980.

[4] OMB, *Guidelines and Discount Rates for Benefit-Cost Analysis of Federal Programs*, Circular A-94 (revised), Washington, D.C., 1992, and Director, Cost Assessment and Program Evaluation, "Economic Analysis for Decision-Making," DoD Instruction 7041.03, Washington, D.C.: Department of Defense, September 9, 2015.

The approach described here views each COA in a product support BCA as a distinct project with an associated sequence of annual flows of benefits (nonfinancial mission and business impacts) and costs (financial mission and business impacts); examines how this sequence might differ under alternative assumptions (sensitivities) and in alternative futures; monetizes the flows in this sequence; and calculates information on the net present value (NPV) of each COA. Commercial companies routinely use this standard approach to compare business alternatives.[5] We believe that this approach offers three benefits to support the decisions that senior government leaders must make when they receive the findings of a product support BCA.

First, it uses dollar measures of value rather than measures of value based on abstract scoring and weighting. This allows us to avoid many of the problems that DoD BCA guidance associates with scores and weights, including explaining (1) the practical meaning of abstract scoring scales, (2) the differences in the policy relevance of different factors scored, and (3) the normalization used to make systems of scores and weights compatible.[6] Senior leaders routinely use dollar values to inform choices among alternatives, especially in the context of the planning, programming, and budgeting processes.

Second, it offers a natural way to integrate thinking about risks with thinking about benefits and costs and uses standard project evaluation tools used throughout the government and private industry.

Third, it also offers a natural way to integrate COA implementation challenges with information about COA benefits and costs. We have found repeatedly that senior leaders have limited interest in a COA that the government will likely have difficulty implementing. The approach described here informs them directly about how implementation difficulties affect the value of a COA.

We developed this approach in the context of conducting the 2014 F-22 Product Support BCA for the Air Force and use concrete examples from that BCA throughout to illustrate elements of the approach. But this approach can be applied in other product support BCAs in the future. We direct those seeking details on the F-22 BCA to documents that describe that analysis in detail.[7] This document emphasizes methodology more than findings relevant to F-22 sustainment.

---

[5] See, for example, Harry F. Campbell and Richard P. C. Brown, *Benefit-Cost Analysis: Financial and Economic Appraisal Using Spreadsheets*, Cambridge, UK: Cambridge University Press, 2003, and Matthew D. Adler and Eric A. Posner, eds., *Cost-Benefit Analysis: Economic, Philosophical, and Legal Perspectives*, Chicago, Ill.: University of Chicago Press, 2001.

[6] These challenges appear in the following sections of Office of the Principal Deputy Assistant Secretary of Defense for Logistics and Materiel Readiness, 2011: 4.4.3.1, "Quantitative and Qualitative Values"; 4.4.3.2, "Scoring and Weighting"; 4.4.3.3, "Quantifying Qualitative Values"; 4.4.3.4, "Normalization"; and 4.4.3.5, "Rank Ordering/Prioritization."

[7] Details on the F-22 Product Support BCA are not available to the general public. Please contact Michael Boito (boito@rand.org) or Kristin Lynch (lynch@rand.org) at the RAND Corporation for information on what materials can be shared.

Chapter Two outlines the approach and explains the potential costs and benefits as a set of cash flows. Chapter Three explains how we identify a short list of risk drivers that we can use to characterize the nature of uncertainty inherent in any COA element. Chapter Four explains how we aggregate information on relevant risk drivers into a single measure of the likelihood of success in any particular year during the implementation of an element of a COA. Chapter Five explains how we structure the collection of data on individual risk drivers and their relationships to elicit professional judgment from subject-matter experts (SMEs) on the uncertainty associated with these risk drivers. Chapter Six explains how we use Monte Carlo analysis to combine this measure, together with information about the potential cash flows associated with the elements of a COA, to generate a subjective probability distribution for each element in a COA and for the COA as a whole. Chapter Seven explains how we can extract information from this distribution and help decisionmakers use this information to choose which COA best matches their preferences. Two appendixes provide additional detail on the workshops used to elicit risk assessments.

2. Assessing the NPV of a COA Element

This chapter provides an overview of the approach. We start by breaking a COA into elements compatible with the level of detail we have in the data. For each COA element, we ultimately translate all measurements related to performance, cost, and risk into dollar terms. This translation allows us to describe a COA element as a series of cash flows over time. These cash flows depend on potential costs and benefits associated with the COA element and on the likelihood that the Air Force can achieve the benefits if it tries to implement the COA element. We treat these cash flows as stochastic and use Monte Carlo analysis to generate many alternative patterns of future cash flows for each COA element. Compiling information on the NPVs associated with each of these patterns of cash flows, we construct a subjective probability distribution that describes the likelihood that the Air Force will realize various levels of NPV if it tries to implement the COA element.

Components of the Approach

To get started, it is useful to understand the definition of a COA element; what the baseline COA is and how we use it when discussing new COAs; how we characterize performance, costs, and risk in dollar terms; and the measure of NPV for a COA element in one potential future. Taken together, these items underlie everything described below.

COA Elements

A COA can comprise a set of changes that differ from one another in important ways. For example, one change might insource the provision of a labor-intensive activity or the management of a specific class of materiel from a contractor to the government. Another in the same COA might change the internal processes that a contractor uses to produce its services in exchange for a change in contractual terms. The COA might further change the geographical location of a contractor or government activity. Each change involves different kinds of potential changes in costs and benefits as measured relative to a baseline. For example, insourcing a labor-intensive activity would require the government to hire, train, and retain additional personnel. If the government can do this, it can reduce its costs if government personnel cost less than contractor personnel with similar skills. Therefore, cost reduction is a potential benefit of this insourcing change. Insourcing the management of materiel might not allow similar savings for labor costs, but could potentially allow the government to avoid paying the contractor a materiel-related fee. This is a different type of change, which would have different costs and benefits.

Each change also faces different uncertainties. When insourcing a labor-intensive activity, the government's ability to hire, train, and retain relevant personnel is crucial. When it insources materiel management, the risks associated with personnel are not as important, but the government's ability to manage and evaluate information on this new materiel becomes important. When such differences exist and we have detailed enough data to treat such changes separately, we do so. When we treat various changes within a COA separately, we refer to each of these changes as a "COA element." For simplicity, the remainder of this chapter speaks only of a COA, implicitly assuming that we are assessing only one element within that COA. The chapters that follow pay closer attention to details relevant to individual COA elements and so give more attention to how COAs and their elements differ.

Difference Between a New COA and the Baseline COA

Any BCA contains a baseline COA—an alternative in which the government chooses not to change any policies, practices, or procedures—and a set of new COAs, in which one or more policies, practices, or procedures change. The baseline COA itself is not static throughout the period of analysis, as circumstances will always change with the passage of time. It simply captures the effect over the period of analysis if the government does not implement a new COA.

A BCA seeks to advise senior leaders on how to choose between any two COAs. With that in mind, the BCA should always focus on the *differences* in benefits, costs, and risks associated with any two COAs. In the discussion below, we discuss the benefits, costs, or risks associated with any new COA. Unless qualified, such statements will always refer to how benefits, costs, and risks in the new COA *differ* from those in the baseline COA.

A BCA need not address absolute levels of benefits, costs, or risks in any particular COA. Consider an example. In any year, the baseline COA and any new COA are likely to include total costs for training and personnel compensation. We do not need to know the actual total costs to conduct the BCA. We only need to know how the costs of training or personnel compensation in the new COA *differ* from those in the baseline COA.

A senior leader can use such a measure to compare any new COA to the baseline COA. But such measures also enable the leader to compare new COAs. Subtracting the NPV of one from that of another provides a useful measure of the difference between the two COAs. Because the baseline COA is always the same, it washes out when a difference is taken between two new COAs.[1]

Standard benefit-cost analysis of alternative projects uses this perspective. It examines the world with and without each project. It then uses information on how the world changes when each project occurs to compare projects against the baseline and against each other.

---

[1] This applies exactly in the sequence of annual flows associated with any one future. The (1) difference in expected value of these annual flows for two COAs need not equal the (2) expected value of the difference in the annual flows for the two COAs. When comparing two new COAs across many potential futures, the second difference is more policy-relevant than the first. We need to keep this caveat in mind whenever looking across many potential futures.

Measuring Performance, Cost, and Risk in Dollar Terms

To treat a COA as a project in a standard benefit-cost analysis, we must measure associated performance, costs, and risks in dollar terms (particularly constant dollar terms). We start by identifying the primary attribute of performance associated with the BCA. In a product support BCA for a major weapon system, the primary attribute is likely to be assurance that the weapon system can perform its missions. Typically, secondary attributes of performance interest DoD only after operational commanders are assured that the weapon system can perform as expected.

For example, senior officials in the Air Force told us that the most important attribute of *performance* for the F-22 is the operational availability of the fleet. The ground rules for the F-22 product support BCA reinforced this perspective by requiring that any new COA sustain the calendar year (CY) 2013 level of operational availability over the entire period of analysis, from fiscal year (FY) 2018 to FY 2033. Other attributes of performance might include fleet agility during deployment, ability to insert new capabilities, and development of organic capabilities relevant to future weapon systems, but senior leaders consider these to be secondary.

To ensure that each new COA held operational availability at the dictated level in the F-22 BCA, we identified actions that the Air Force had to take to sustain such availability in each new COA. If a new COA insourced workload, for example, we considered the technical data that the Air Force needed to ensure access and the new personnel that the Air Force needed to hire, train, and retain. To further ensure the required level of operational availability, our analysis retained contractor personnel in place while new government personnel trained and assumed their roles alongside their contractor counterparts. If a new COA involved a new approach to contracting for product support, we considered the actions that the Air Force had to take to design and implement such a new approach. We assigned the costs of such actions to each new COA as appropriate. Where inherent differences in new COAs were likely to lead to differences in operational availability, we adjusted the inventory that the Air Force used to support the F-22 fleet so that each COA under consideration achieved the same level of operational availability in the fleet.[2] If a new COA degraded availability relative to that in CY 2013, we added inventory, which imposed a dollar cost on the Air Force. If a new COA enhanced availability relative to that in CY 2013, we removed inventory, reducing dollar costs to the Air Force over the long run.

Following this adjustment, operational availability was no longer a discriminating factor relevant to a product support BCA. Any differences in operational availability relevant to discriminating among COAs were captured in the NPVs of new COAs to be compared. (We

---

[2] To do this, we applied the Aircraft Sustainability Model (ASM), a tool that the Air Force often uses to link availability levels to cost. This model provided a widely used and understood technique to translate a performance measurement into a dollar value. For details on the tool, see F. Michael Slay and Randall M. King, *Prototype Aircraft Sustainability Model*, Report AF601R2, McLean, Va.: Logistics Management Institute, 1987; Craig C. Sherbrooke, *Optimal Inventory Modeling of Systems: Multi-Echelon Techniques*, New York: John Wiley and Sons, 1992; and *ASM® Sparing Model*, McLean, Va.: Logistics Management Institute, 2012.

explain how we reflect other, secondary attributes of performance in the analysis in the next section.)

Cost enters the analysis in two ways. The first is through actions that occur early in a new COA, such as upfront investments, with the expectation that the new COA will generate benefits later. The second is through the generation of savings later in the new COA. For example, insourcing an activity can ultimately save the government money if government personnel cost less than contractor personnel with comparable skills. Moving an activity to a lower-cost location can ultimately reduce costs, mainly through changes in labor compensation. Changes in contractor processes can ultimately reduce contractor charges. Other savings might come from the following sources:

- improving system reliability
- improving vendor base management
- substituting government for contractor material management
- changing government-contractor interaction.

We define *risk* in terms of the government's ability to realize the maximum net benefits from implementing a COA. Some expect the government to have no difficulty implementing a new COA, allowing a high probability that it will achieve large net benefits. When someone views a new COA this way, we say that the person believes the new COA involves little risk. Others believe that the government is unlikely to achieve the full net benefits of a new COA. When someone views a new COA this way, we say that the person believes the new COA involves more risk.

Defined in this way, risk enters this analysis in two potential ways. The first involves the initial cost of ensuring that the government can sustain an acceptable level of system performance (in the F-22's case, operational availability) under the COA through appropriate investments in data, personnel training, and any other necessary resources. The higher the risk associated with a new COA, the more the government might have to invest to sustain system performance. The second involves the net savings that the government ultimately achieves when it implements a new COA. The higher the risk associated with a new COA, (1) the longer it takes for the government to implement it, delaying net benefits and protracting implementation costs, and (2) the smaller share of potential net benefits the government realizes by the end of the period of analysis. That is, higher assessed levels of risk can enter the analysis either by increasing the cost of implementing a new COA or reducing the net benefits the government ultimately realizes through the new COA. Either way, increased risk reduces the NPV that the analysis associates with a new COA.

Risks to the government's ability to implement a COA could arise if any of the following occur:

- The government cannot execute a new plan.
- The government and contractor cannot agree to transfer technical data.

- The government lacks tools to manage technical data.
- The government cannot attract, train, and retain personnel.
- The government cannot cost-effectively manage the vendor base.

The approach outlined above enables us to treat all of these aspects—performance, cost, and risk—in dollar terms. As a result, we do not need to explain the relative importance of different analytical factors to senior decisionmakers. The dollar values we use convey the needed information on relative value.

More generally, this approach allows us to combine information on the performance, cost, and risk associated with any new COA into a single dollar measure of net value in each year during the period of analysis. We then use these dollar values for annual flows to conduct a standard project evaluation. That evaluation links an NPV to the sequence of annual flows associated with each new COA.

Secondary Attributes of Performance

We expect the warfighter's primary goal, however defined, to dominate other attributes of performance. However, we do not want to preclude the possibility that other secondary attributes of performance could shape the preferred COA. To shape the outcome of a BCA, the value of a secondary attribute would have to vary across COAs. By definition, it would have to be favorable enough to change the outcome of the BCA if the warfighter's primary goal were the only attribute of performance considered in the BCA.

Suppose that, treating the warfighter's primary goal as the only attribute of performance in a BCA, this approach yields a recommendation that the government pursue COA One. Under this approach, COA One is, by definition, the COA with the highest NPV that achieves the warfighter's target goal. Now add a second attribute of performance that varies in value across COAs. Decisionmakers could prefer a "best value" COA Two with a higher value of this second attribute than COA One, if they were willing to pay the difference in the values of NPV to benefit from the level of the second attribute available in COA Two.

For example, suppose decisionmakers cared strongly about insourcing an activity associated with F-22 fleet sustainment, and suppose that this approach indicated that an insourcing COA that maintains fleet operational availability is very likely to have a lower NPV than any other new COA that maintains fleet operational availability. Then the differences in NPV among new COAs that this approach estimates can help decisionmakers understand how much they must be willing to pay to pursue insourcing rather than preferring the COAs with higher NPVs. That is, this approach does not reject decisionmakers' preference for a new COA with a lower NPV. Rather, it clarifies what the decisionmakers must be willing to pay to pursue the benefits to the government that they associate with the new COA with a lower NPV.

NPV for a COA

In sum, the approach applied here gives decisionmakers information that measures COA performance, cost, and risk with a single common currency—the NPV of annual flows of costs and benefits over the period stated in constant dollars.

In any potential future, we represent each new COA as a series of cash flows. We use the OMB's prescribed discount rate to aggregate each series of cash flows in a potential future into a single NPV for that potential future.[3] This NPV is the elementary unit of analysis that we use throughout the analysis. We can use it to do the following:

- make a pairwise comparison between corresponding elements in any two new COAs in any potential future
- build a subjective distribution of NPVs for any one new COA element, relative to the baseline COA, across all potential futures
- build a subjective distribution of a pairwise comparison between corresponding elements in any two new COAs across all potential futures.

In each application, the NPVs reflect differences in cash flows between each new COA and the baseline COA and, equivalently given the linear definition of NPV, differences between the NPVs of each new COA and the baseline COA. As a practical matter, however, the approach never generates the information required to make this last calculation directly.

From a Series of Cash Flows to a Subjective Probability Distribution of NPV

As noted above, we present each new COA in terms of a series of cash flows. We start with information like that shown in Figure 2.1. Time is on the horizontal axis. In this example, time runs over the period of analysis specified in the terms of reference for the F-22 product support BCA, from FY 2018 to FY 2033. Notional annual flows are shown on the vertical axis in constant FY 2015 dollars. To begin, we gather information on all the costs likely to be associated with initiating a new COA, relative to those in the baseline COA. Figure 2.1 displays these notional values in a simplified form as constant annual negative cash flows that continue for several years as the government initiates the new COA. We also gather information on the maximum potential annual net benefits that the government might garner from the new COA, relative to those in the baseline COA. Figure 2.1 shows these in a simplified form as a series of constant positive cash flows from the date at which full implementation occurs to the end of the period of analysis.

---

[3] OMB, "Discount Rates for Cost-Effectiveness, Lease Purchase, and Related Analyses, Appendix C (revised)," December 2014, in *Guidelines and Discount Rates for Benefit-Cost Analysis of Federal Programs*, Circular A-94 (revised), Washington, D.C., 1992.

Figure 2.1. Notional Series of Cash Flows for a New COA

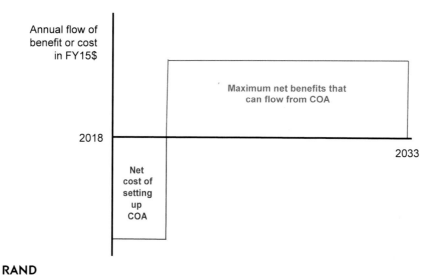

This series of potential future cash flows will be familiar to anyone who has conducted quantitative project evaluation. These cash flows capture the costs and savings that we associate with performance and cost.

Then we adjust these potential values to reflect the shares of these that the government will encounter if it pursues this new COA. The red bars in Figure 2.2 indicate that, in one potential future, the government encounters the full potential costs of initiation in the first year, but that the level of these costs falls in each of the next two years. In the fourth year, annual net cash flow values turn positive and continue to increase for several years. But they never rise as high as the full potential net benefit level that the government might have achieved. These adjustments in the potential cost and benefit levels capture the effect of risk in this new COA.

10

Figure 2.2. Notional Cash Flows for a New COA in One Potential Future

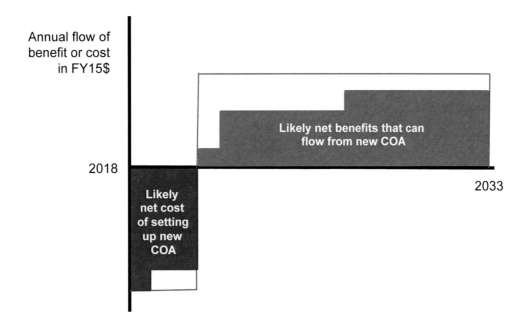

The adjustment for risk is less traditional than the simple presentation of potential net costs and net benefits. Chapters Three to Six explain in detail how this approach determines what the adjustment should be for a particular new COA in any potential future. For now, note simply that the adjustment differs by year in any potential future. Generally speaking, this approach tends to generate adjustments that enable success in implementation to increase over time but often fall short of 100 percent success.

We use Monte Carlo simulation to generate potential futures for each new COA. Monte Carlo methods are often used in project evaluation to

> build up random scenarios which are consistent with the analyst's key assumptions about risk. [Such] a risk analysis application utilises a wealth of information, be it in the form of objective data or expert opinion, to quantitatively describe the uncertainty surrounding the key project variables as probability distributions, and to calculate in a consistent manner its possible impact on the expected return of the project . . . . [Such] use of risk analysis in investment appraisal carries sensitivity and scenario analyses through to their logical conclusion.[4]

Each future can be described or defined by a series of cash flows. Figure 2.3 displays three series to illustrate the approach. That shown in red assumes that the full potential costs and benefits occur as shown in Figure 2.1. That shown in purple represents the series shown in Figure 2.2. The brown locus shows a third possibility. Each locus yields a single NPV for the potential future shown.

---

[4] Savvakis Savvides, "Risk Analysis in Investment Approach," *Project Appraisal*, Vol. 9, No. 1, March 1994, p. 3. See also Marius Holtan, "Using Simulation to Calculate the NPV of a Project," InvestmentScience.com, May 31, 2002.

Figure 2.3. Cash Flows in Three Potential Futures for One New COA

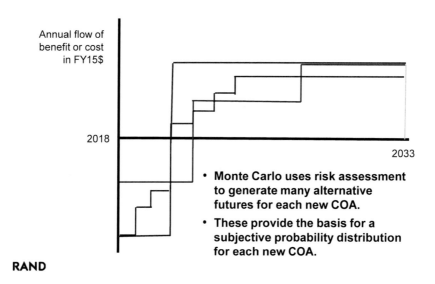

RAND

When we gather all of these NPVs from all potential futures and plot their frequency, we get a distribution similar to that shown in Figure 2.4. The NPV for potential futures appears on the horizontal axis. The probability frequency for these futures appears on the vertical axis. Frequencies shown in red represent negative NPVs. Those shown in green show positive NPVs. By definition, the areas shown in red and green together sum to one. In effect, for each new COA, this distribution brings together in one place all the information we have on the costs of achieving the primary performance goal, investment costs for preparing for the new COA, savings generated through the new COA, and risks relevant to these costs and savings.

Figure 2.4. Subjective Probability Distribution Across All Potential Futures for One New COA

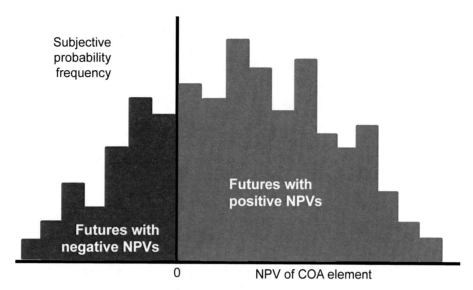

Distribution provides basis for various measures of COA desirability; e.g.,
  • subjective confidence interval
  • subjective expected value
  • subjective probability of breaking even

By itself, this distribution could be all a decisionmaker needs to assess a new COA. In all likelihood, however, a decisionmaker will prefer some summary expression of the information in Figure 2.4. Here are three possibilities:

  • **Probability of loss.** A decisionmaker with a high aversion to loss might want to use the area shown in red to compare competing COAs.[5] If this is the only thing the decisionmaker cares about, such a decisionmaker would prefer the new COA with the smallest red area. In this case, the decisionmaker only needs information on the area in red, which can be summarized in a single number.

  • **Expected value.** If a decisionmaker has no aversion to loss or risk and is willing to accept losses in some programs with the expectation that gains in other programs will more than offset these in aggregate, the theory of rational decisionmaking recommends that the decisionmaker use the expected value of dollar NPV associated with the distribution shown in Figure 2.4.[6] If this is the only thing the decisionmaker cares about,

---

[5] A decisionmaker is loss-averse when any one-dollar loss reduces the decisionmaker's utility and a sense of wellbeing by more than any one-dollar gain increases the decisionmaker's utility. Traditional risk assessment emphasizes the importance of loss averseness by focusing on the likelihood and consequence of loss and preferring options with smaller potential losses regardless of what potential gains they might offer.

[6] A decisionmaker is risk-averse when each additional gain of one dollar yields less utility for the decisionmaker than the last dollar gain did. A decisionmaker who is not loss- or risk-averse by definition gains (loses) the same incremental utility from every dollar gained (lost). Rational decisionmaking calls for a decisionmaker to maximize

such a decisionmaker would prefer the COA element with the highest expected value of dollar NPV. In this case, the decisionmaker only needs information on this one value, which can be stated in a single number.

- **Subjective confidence interval.** A decisionmaker with moderate risk aversion might want to clip areas of equal size off both tails of the distribution and ask for the values of NPVs that achieve such a cut. For example, an 80-percent confidence interval would clip 10 percent from each tail. The interval between the two points would provide an estimate based on subjective judgment of where the NPV for a COA element might lie in 80 percent of the potential futures.[7] Such a decisionmaker would generally prefer a COA element with a narrower interval between these cut points, but would accept a broader one if a large enough difference in expected value offset it. Conversely, such a decisionmaker would accept a COA element with a lower expected value if that COA element offered enough certainty around that lower value.

Decisionmakers exist with priorities that can be characterized in each of these ways (and others). As analysts, our job is not to tell a decisionmaker what her or his priorities should be. Rather, it is to present systematically organized information that can be used to implement priorities. A subjective probability distribution, such as that in Figure 2.4, provides such information. The three measures of performance highlighted are examples of summary statistics that could potentially help decisionmakers with differing priorities use the information shown in such a distribution.

Risk analysis often defines risk in terms of two components: (1) the probability of loss and (2) the magnitude of loss if it occurs.[8] Risk rises as either the probability or the magnitude of loss increases. Such analysis often displays risk in a two-dimensional "risk probability and impact" matrix with five rows to present five progressively increasing levels of probability of loss and five columns to present five progressively increasing levels of magnitude of loss. Such analysis of a COA element would place any particular COA element in a single cell; the farther a cell lies toward the upper right in the matrix, the higher the risk associated with any COA element in the cell.

In a product support BCA, such an approach might report the NPV of a COA element if implementation occurs without any problems and realizes the full net benefits available from the COA element. It might then use a matrix like the one described above to describe the magnitude of a potential shortfall relative to this maximum level of benefit and the probability that this

---

the expected utility. When neither loss- nor risk-averse, the decisionmaker can maximize utility by preferring the COA with the highest expected dollar NPV.

[7] Note that this is not the same kind of confidence interval used in statistical inference. It is not based on drawing a sample from an objective population and using data from the sample to infer attributes of the underlying population. Rather, it seeks to summarize the nature of a person's or group's subjective beliefs about the world. Because it does not involve sampling, there is no sample size to report when providing a subjective confidence interval.

[8] For a discussion of this approach and its application in the context of a risk probability and impact matrix, see Office of the Principal Deputy Assistant Secretary of Defense for Logistics and Materiel Readiness, "Risk Analysis," Section 4.7.1, 2011.

shortfall might occur, each defined qualitatively so that the risk assessment simply places the potential loss in one of the 25 cells of the matrix.

Such analysis treats a COA element as though a single magnitude of loss can occur and that magnitude occurs with a probability that we can state with a single point value. The subjective probability distribution in Figure 2.4 moves beyond this simple description of risk in two ways. First, it shows that many different magnitudes of loss (relative to the maximum potential gain) can occur, each with its own probability. Second, it replaces the qualitative character of the cells in the matrix with numerical assessments of (1) the dollar value of magnitude of loss and (2) the subjective probability associated with each dollar value of magnitude of loss. In this way, it offers a much richer description of what might occur in the future and one better suited to helping decisionmakers, who are in effect making an investment decision, reflect any loss aversion or risk aversion they might harbor in their choice of a preferred investment option.

3. Risk Drivers Relevant to COA Elements

This chapter uses our analysis of the air vehicle in the F-22 product support BCA to illustrate the first stage of a two-stage data collection designed to identify risk drivers relevant to any COA element and understand what characteristics of each risk driver affect the benefits that the government can ultimately realize from pursuing any COA element. Our analysis of the F-22 yielded eight risk drivers that appeared repeatedly in various COA elements. It also yielded a short list of characteristics likely to be important to the construction of a formal model of the risks associated with any COA element. These risk drivers and characteristics are likely to be relevant to assessing risk in future product support BCAs.

The chapter begins with a brief description of aspects of the F-22 air vehicle that help explain why we should expect problems—"risk drivers" or "sources of risk"—to be important to sustainment of the F-22 air vehicle. It briefly weighs two alternative methods to assessing such risk drivers. It then describes how the approach described here applies one of these methods to identify risk drivers to examine in greater detail. It uses one COA element associated with the air vehicle in the F-22 product support BCA to illustrate the approach. Applying this method to many COA elements revealed characteristics of risk drivers that appeared repeatedly. These characteristics framed the formal risk model that we built for the F-22 BCA.

Aspects of the F-22 Air Vehicle Relevant to Our Analysis

The F-22 is a complex aircraft with many unique attributes. It is the first fifth-generation fighter aircraft in the Air Force inventory. The Air Force has procured 187 total operational platforms. This number is relatively small when compared to most other fighter aircraft programs. The F-22 program had a relatively high degree of acquisition concurrency, during which the development and production phases of the program significantly overlapped, creating higher risks than exist in a standard acquisition. To achieve this level of concurrency, from the beginning, the Lockheed Martin–Boeing team developed the aircraft with advanced, integrated design, development, and production software that keeps sustainment processes centralized and comparatively streamlined. The Lockheed Martin–Boeing team developed some of this software exclusively for this aircraft; much of it is proprietary. High levels of concurrency led to a relatively higher number of platform modifications than occur on a per-platform basis on legacy aircraft, such as the F-16. Today, each aircraft has a unique configuration. Sustainment functions differ slightly for each individual aircraft structure, requiring close sustainment management by specific tail number.

From a technological standpoint, the air vehicle contains a high percentage of advanced alloy materials and graphite composites, including in primary load-bearing structures and substructures. It also contains a highly specialized coating to maintain a lower radar signature. For enhanced flight dynamics, it is equipped with 2-D thrust vectoring and supersonic cruise capability (the ability to fly at supersonic speeds without the use of afterburners). From an operational standpoint, given the relatively small number of platforms and the essential role that the F-22 might play in the early phase of conflicts as an enabler for other missions, maintaining F-22 availability is important. Small delays on a small number of aircraft can potentially have a large effect on overall Air Force and joint force readiness. This importance helps explain the heavy emphasis that Air Force officials placed on maintaining availability in the F-22 fleet during the BCA.

Two Alternative Approaches to Identifying Risk Drivers

We considered two ways to identify risk drivers relevant to the F-22 air vehicle:

- a structured general model
- a tailored expert model.

Risk assessors often use the structured general model when examining a system similar to earlier systems, which are likely to be susceptible to similar sources of risk and conducive to similar risk mitigation approaches. Historical experience informs assessors about what sources of risk to emphasize when examining the new system. It enables a closely structured instrument and well-defined rating scales to measure risk. This approach works well when such a structured approach allows an assessor to build constructively on historical experience. But it can understate risk if the new system differs substantially from its historical precedents. When large differences exist, a closely structured analysis can prevent an assessor from discovering risks and/or mitigations not present in the past.

The 2010 F-22 BCA used such a structured, general approach to assess risk.[1] That risk analysis applied a structured risk assessment model that evaluated COA elements in terms of the level or quality of the (1) knowledge, (2) information technology (IT) tools and processes, (3) capabilities, and (4) effectiveness associated with each COA element. The evaluation process organized COA elements around the integrated process teams that the Lockheed Martin–Boeing team used to oversee particular activities relevant to each COA; each element in a COA addressed the activities associated with a different integrated process team. Several SMEs subjectively assessed COA elements in terms of these four criteria. The 2010 F-22 BCA did not discuss unique risk drivers specific to individual COA elements or to what extent these should be represented.

---

[1] Bob Rhea, Steven Hurt, Alan Heckler, Sameer Dohadwala, Hamza Rampurawala, and Rajan Singh, *Recommendation for Long-Term Sustainment of the F-22 Raptor: F-22 Sustainment Business Case Analysis*, Chicago: A. T. Kearney, October 2009.

A tailored expert model does not use a closely structured approach to collect information on risk. Instead, it relies on open-ended discussions with SMEs and builds a new structure as information accumulates. Early interviews help a risk analyst construct an expert model tailored to the risks associated with the weapon system under study. Later interviews add detail and allow the analyst to test elements of the model built on the basis of earlier interviews. Standardized instruments and scales are harder to apply, complicating the comparison of findings to historical norms even as the tailored expert model matures.

In the context of the second F-22 BCA, the high degree of concurrency in the development and production of the F-22 air vehicle and the combination of advanced technologies incorporated within the platform suggest that the risk drivers relevant to COA elements in this new product support BCA would likely differ from those that have been important to the product support for previous legacy aircraft. As a result, generalized methods and criteria used to assess COA elements may not serve as adequate precedents for evaluating risk relevant to sustaining the F-22 air vehicle. New sources of risks and unique criteria could be important. This suggests that a tailored expert model created specifically for this system should be used.[2] That is, as we sought insight into the likelihood that the Air Force could realize the full net benefits potentially available from a COA element, specific sources of risk might require special attention. New criteria relevant to the sources of risk might be appropriate. With this in mind, we focused on the second form of risk analysis.

Tailored Expert Model

As in Chapter Two, our approach begins with a specific COA element. We asked knowledgeable SMEs, with a variety of different interests, to identify sources of risk relevant to this COA element. The initial queries were relatively open-ended and unstructured. In the F-22 product support BCA, we sent open-ended questionnaires to relevant SMEs early in the research process to provide an initial identification of key risk drivers. The SMEs included representatives of the Air Force system program office and SMEs at field locations where the Air Force currently operates and sustains F-22s. They also included representatives of the primary contractors that sustain the F-22, including Lockheed Martin, Boeing, and Pratt & Whitney. We received responses from all these organizations. We used information in these responses to identify an initial set of risk drivers. We followed up with in-person interviews to elicit more-detailed information on risk drivers. In our approach, interviews covered historical information and additional expert information on the identity and nature of current risk drivers.

---

[2] We derived the approach presented here from one described in M. Granger Morgan, Baruch Fischhoff, Ann Bostrom, and Cynthia J. Atman, *Risk Communication: A Mental Models Approach*, Cambridge, UK: Cambridge University Press, 2002, reprinted 2011.

The approach described here organizes the information collected in this way into a formal tailored expert model for each COA element. This model organizes what we heard into a transparent framework that shows how sources of risk relate to one another and how they work together to affect the government's ability to realize the full potential net benefit associated with each COA element. This model highlights causal links among risk drivers and helps us identify gaps in our understanding that help us focus ongoing queries increasingly on specific issues relevant to the COA element. As the model for a COA element becomes increasingly refined—that is, as interviewees become less and less able to add risk drivers to help us understand problems that have arisen in the past—we summarize the key information in each model in a risk influence diagram.

Figure 3.1 provides a diagram of the risk drivers that our data collection associated with a COA element focused on the F-22 air vehicle Aircraft Integrity Program (AIP), an activity present in most Air Force aircraft programs. It gathers information on a maturing, then aging fleet to help understand where specific future sustainment and upgrade actions are likely to be most cost effective. The COA in question considered transferring a portion of this program from provision by the Lockheed Martin–Boeing team to organic provision within the Air Force F-22 program.

Solid ovals in Figure 3.1 display risk drivers that the first round of interviews with SMEs identified. According to these SMEs, the Air Force's ability to acquire formal technical data packages affects its ability to access current technical data that it needs to conduct AIP tasks. Together with access to appropriate analytic and data management tools and dependence on hiring and training skilled staff, the Air Force's access to current technical data affects its ability to meet AIP objectives. Any compromise in the Air Force's ability to meet these objectives could compromise the Air Force's ability to realize the full benefits of bringing AIP activities and capabilities in house.

The second round of interviews offered a more nuanced and detailed set of risk drivers and links among them. These drivers are represented in Figure 3.1 by dashed ovals. SMEs stated that the Air Force's ability to access current technical data depends not only on its current ability to acquire formal technical data packages but also on its access to evolving technical data. Because technical data relevant to F-22 sustainment change almost daily, acquiring a snapshot of these data is not enough. SMEs questioned the Air Force's understanding of fifth-generation technology and its resulting ability to develop personnel skills relevant to ongoing F-22 sustainment; even if the Air Force could acquire the right people, it could have difficulty training them and keeping them current on fifth-generation technical issues. SMEs questioned the ability of Air Force personnel to react in real time to unexpected problems. With a limited ability to react to such shortfalls, the Air Force could have difficulty keeping its sustainment workforce current. Finally, the second round of interviews added a potential problem: managing changes associated with low observable (LO) technology. Taken together, these additional risk drivers raised questions about whether the Air Force could fully realize the net benefits it expects from insourcing some AIP activities.[3]

Figure 3.1. Risk Influence Diagram for Insourcing Portions of the F-22 AIP Program

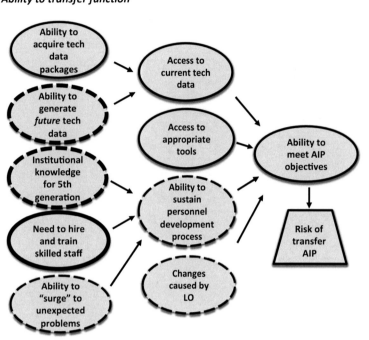

Ability to transfer function

- Initial interviews identified risk drivers (solid ovals).

- Second-round interviews added drivers (dashed ovals).

- We elicited details on risks (primarily associated with ovals in central column).

The tailored expert models that we built and the associated risk influence diagrams gave us three important insights.

1. They identified a field of eight risk drivers that arose repeatedly in discussions about the different COA elements.
2. They raised a set of issues that would clearly be important when we looked at all the risk drivers associated with one COA element and sought a summary risk assessment that considered the effects of all risk drivers.
3. They provided enough detail about the risks associated with F-22 sustainment to build a formal, structured model of this risk.

However, we would need to collect information on relevant risks in a more systematic way to make defensible comparisons of competing COA and COA elements. We consider each of these insights in turn.

---

[3] In particular, this approach to assessing risks associated with the COAs proposed under this BCA revealed risks not identified or assessed in the 2010 F-22 Sustainment BCA.

Potential Areas of Risk or Risk Drivers Relevant to F-22 Sustainment

This initial analytic effort identified eight risk drivers that are collectively relevant to the COA elements addressed in the F-22 Product Support BCA. Different risk drivers applied to each COA element that we examined. The tailored expert models and associated risk influence diagrams developed for a future product support BCA would likely yield some of these risk drivers, but would also probably identify others. We offer here the eight risk drivers identified in our F-22 BCA to suggest the kinds of risk drivers that might arise in future product support BCAs. The drivers on this list also play an important role in Chapter Four.

Difficulty Hiring, Training, and Retaining Relevant Personnel

This risk driver involves the government's ability to obtain, sustain, and retain personnel with the skills needed to transfer functions to the government. Given that the F-22 is a highly specialized platform, a personnel risk may be associated with the transfer of specialized functions. Subfactors include the ability to attract and retain qualified personnel, to address salary differentials, and to train personnel. An on-the-job training program with Lockheed Martin, Boeing, or Pratt & Whitney personnel is envisioned for any specialized functions as required.

Difficulty Accessing Relevant Technical Data

This risk driver involves the government's ability to access and use technical data packages within each of the transferring functions—unless, however, these technical data packages will be provided by the Lockheed Martin–Boeing team or Pratt & Whitney at no additional cost. The F-22 and F119 were designed and produced using a state-of-the-art paperless system. Their technical data are currently housed within this proprietary system, and the government will have difficulty transferring functions in house if it cannot access or transfer existing and future data. Since the F-22 was produced with a high level of concurrency, the number of modifications and changes to technical data are unusually high relative to other aircraft in the Air Force inventory.

Difficulty Accessing Relevant IT and Proprietary Tools

This risk driver entails the government's ability to access proprietary tools associated with current F-22 sustainment functions and to implement similar (or identical) IT. Specifically, the F-22 uses specialized tools to facilitate the execution of various functions. There is risk associated with the transfer of these functions, given that the government does not have access nor does it currently operate these tools. The government needs to either have comparable tools or be able to arrange access to Lockheed Martin, Boeing, or Pratt & Whitney tools, for example, through a licensing agreement or other contract vehicle.

Difficulty Accessing Information Software Systems

This risk driver involves the government's ability to transfer information software systems from the Lockheed Martin–Boeing team or Pratt & Whitney. Specifically, the Lockheed Martin–Boeing team and Pratt & Whitney use proprietary forecasting tools and track the percentage of the F-22 or F119 fleet that has each item or piece of equipment to predict demand. These systems are different from those used in organic supply system processes.

Difficulty Developing or Executing a New Contracting Vehicle

This risk driver entails the government's ability to develop and execute a long-term contract (for example, five-year base with a five-year option) that would be optimal for providing incentives to the Lockheed Martin–Boeing team or Pratt & Whitney to invest in cost-saving measures.

Difficulty Managing Institutional Knowledge Relevant to the F-22

This risk driver is associated with the government's competency and depth of knowledge in managing a specialized fifth-generation aircraft, particularly in addressing unanticipated problems. As with personnel, technical data, and tools, given the specialization of the F-22 air vehicle and engine, the Lockheed Martin–Boeing team and Pratt & Whitney have developed F-22–specific institutional knowledge in managing these programs. While the government could overcome its current lack of data and tools over time, it may take more time to develop the base of knowledge to use the information in a productive and efficient manner. This includes where and who to go to in order to address unexpected problems.

Difficulty Acquiring Knowledge About the Vendor Base or Managing Relationships

This risk driver involves the government's ability to take on specific supply functions. Today, the government does not have the same long-term agreements and relationships with vendors that the Lockheed Martin–Boeing team and Pratt & Whitney have to ensure a robust supply base.

Difficulty Ensuring Product Support Processes Comparable to Those in Place Before the BCA

This risk driver involves the government's ability to provide a product support process comparable to the Lockheed Martin–Boeing team's and Pratt & Whitney's current processes. As with institutional knowledge, this risk driver focuses on the product support processes themselves.

Translating Assessments of Individual Risk Drivers into a Summary Assessment for a COA Element

The construction of the tailored expert models and associated risk influence diagrams revealed several judgments that an analyst would have to make to integrate assessments of individual risk drivers into a summary assessment of risk associated with each COA element. The judgments proved to be similar enough across COA elements to suggest a common approach to assessing risk for each COA element in the F-22 Product Support BCA. This approach is likely to be useful in future BCAs as well.

Risk Drivers Relevant to a COA Element

Which risk drivers listed above are *relevant* to any particular COA element? Are any additional risk drivers relevant?

Probability That Problems Associated with Any Risk Driver Can Be Mitigated

For each risk driver relevant to a COA element, what is the *probability* that the government can undertake changes that ensure that it can mitigate problems associated with that risk driver well enough to ensure the success of the COA element? In the AIP example shown in Figure 3.1 above, what is the probability that the government can undertake changes that ensure that the government can access relevant technical data? What is the probability that the government can undertake changes that ensure that the government can access relevant tools, maintain a workforce with suitable skills, or handle any new problems with LO technology that arise in the future? It is intuitive to expect that, all else equal, increasing the probability that the government mitigates problems associated with a risk driver relevant to a COA element should increase the probability of successfully implementing that COA element.

Criticality and Substitutability of Risk Drivers

How *critical* is the mitigation of problems induced by any one risk driver relevant to a COA element to successful implementation of the COA element? For example, several interviewees suggested that, in the absence of access to technical data relevant to a COA element, the COA element simply cannot be implemented. Mitigating problems associated with other risk drivers is irrelevant until the government has suitable access to relevant technical data. When this occurs for any risk driver, we call that risk driver critical to successful implementation of a COA element.

In other cases, interviewees indicated that mitigating problems associated with one risk driver could also mitigate problems with another, even if that risk driver still presented significant shortfalls. For example, it is possible that, if the government has access to personnel with the right skills, it can implement a COA element successfully without access to all the tools it might hope to have. In such a case, the government can use higher-quality personnel to overcome shortfalls with regard to high-quality tools; the government can, in effect *substitute* higher-quality personnel for high-quality tools. When such substitution can occur, the government can increase the probability of successful implementation of a COA element by adding higher-quality personnel, even if it has not fully mitigated a shortage of high-quality tools. Such an argument is analogous to saying that a kitchen can increase the likelihood to serving a high-quality omelet by hiring a high-quality chef, even if it does not have a pan optimized to make omelets. It is intuitive to expect that, all else equal, increasing the substitutability between risk drivers relevant to a COA element should increase the probability of successfully implementing the COA element.

When such substitution is feasible, the government can increase the probability of successful implementation in a variety of ways. When that is possible, what is the *relative importance* of mitigating the various risk drivers that are relevant to a COA element? Put another way, in the presence of shortfalls in several substitutable risk drivers, which should the government address first to have the most cost-effective outcome? It is intuitive to expect that, all else being equal, increasing the relative importance of a risk substitutable driver relevant to a COA element, with associated problems that the government has mitigated, should increase the probability of successfully implementing the COA element.[4]

Integrating the Factors Above to Assess the Probability of Successful Implementation

If we consider all the factors above, an assessment of the probability of successfully implementing a single COA element might depend on assessments for each of eight risk drivers of the (1) probability of mitigating problems associated with the risk driver, (2) whether mitigation of problems associated with the risk driver is critical to the success of the COA element or whether the government can resolve these problems by substituting mitigations of problems associated with another risk driver, and (3) the relative importance of each substitutable risk driver to the successful implementation of the COA element—as many as 24 assessments to understand the risks associated with any single COA element. It would be difficult for even an expert to do this in any heuristic way. It would be even more difficult to document the basis for the expert's judgment. Some systematic mechanism would be required to bring all these assessments together in an internally consistent, transparent way that could be shown not to favor, even inadvertently, any one COA element over a competing COA element.

Chapter Four describes an "aggregation function" that we identified to bring together all the considerations above as simply and transparently as possible.

---

[4] Of course, any risk driver can only be *relatively* important to the successful implementation of a COA element. So it would be technically more nearly correct to say the following: All else equal, raising the importance of a risk driver that the government has a good chance of managing *at the expense of* a risk driver that the government will not be able to manage as well should increase the probability of successfully implementing the COA element.

Collecting Data on Risk to Support Defensible Comparisons Between COA Elements

We could implement the mechanism called for above, of course, only if we had clearly defined inputs on all the relevant factors, then collected input in a transparent way that ensured that risk assessment compared all competing COA elements fairly. As a practical matter, we discovered that we could not simply define parameters and then collect values of these parameters from SMEs. Rather, formal risk assessment methods have found that SMEs can express their beliefs more reliably when parameters are defined in a way that favors their intuition. We also concluded that we could not assemble the parameter values we needed from the risk assessment we had conducted to build the tailored expert models and associated influence diagrams. We would need a new data collection that was significantly less open-ended to ensure that SMEs focused on a short list of important issues and addressed them consistently. Chapter Five describes this second stage of data collection.

4. Aggregating Information on Different Risk Drivers Relevant to a COA Element

Our development of tailored expert models made it clear that we needed a systematic way to translate (1) assessments of the degree of success in controlling individual sources of risk or risk drivers into (2) a measure of their joint effects on the degree of success realized in a particular COA element. This chapter describes an aggregation function that does just that. In particular, it does it in a way that instantiates the intuitions we have about how features of risk drivers relevant to a COA element should affect the probability of successfully implementing a COA element.

Let p_i be the subjective value of the degree of success in controlling the ith risk driver. Let p_T be the subjective value of their total, joint effects on the degree of success realized in a particular COA element. We seek an aggregation function with the following properties:

$$0 \leq p_i \leq 1 \text{ and } 0 \leq p_T \leq 1.$$

(Eq. 4.1)

$$p_T = 0 \text{ when all } p_i = 0, \text{ and } p_T = 1 \text{ when all } p_i = 1.$$

(Eq. 4.2)

$$\partial p_T/\partial p_i \geq 0, \ \partial p_T/\partial r_i \geq 0, \text{ and } \partial p_T/\partial s \geq 0.$$

(Eq. 4.3)

for r_i, a measure of the relative importance of the ith risk driver, and s, a measure of the ability to substitute among risk drivers. If it is impossible to increase the value of p_T by raising the value of any one p_i without raising the others proportionately, then $s = 0$. If $s > 0$, an increase in the value of any p_i can increase p_T without simultaneously increasing the value of p_i for any other risk driver. When this is true, we can in principle make up for a difficulty in controlling one risk driver by improving our control of another—in effect, *substituting* control of one risk driver for control of another—hence the ability to substitute among risk drivers.

Ideally, we would prefer an aggregation function that allows s_{ij} to differ for any pair of risk drivers i and j. The transcendental logarithmic production function used widely in economics to relate inputs to outputs—in effect, to aggregate levels of inputs into levels of outputs—would allow that.[1] In our setting, we believe it would be unrealistic to identify such pairwise values of

---

[1] K. J. Arrow, H. B. Chenery, B. S. Minhas, and R. M. Solow, "Capital-Labor Substitution and Economic Efficiency," *Review of Economics and Statistics*, Vol. 43, No. 3, August 1961, pp. 225–250. A formal *elasticity of substitution* between two inputs, σ_{ij}, is the elasticity of the ratio of the two inputs with respect to the ratio of their

s_{ij} for the risk drivers we are using. No empirically based literature is available to inform any such choice of values. And, in part because of this, we think it would be unrealistic to use SMEs to develop such detailed information.

Economists use a related type of aggregation function, called a *constant elasticity of substitution production function*, which uses one single value of σ to characterize substitutability among all inputs.[2] Unfortunately, that function is not well defined when any input takes a value of 0. Therefore, we cannot apply it in our setting. In effect, no aggregation function exists that allows us to apply any single global value of σ for all risk drivers for which $0 \le p_i \le 1$.

To meet our needs, we seek an aggregation function that enables us to reflect two different levels of substitution among risk drivers. We use a nested function to do this. Some risk drivers may be so critical to controlling the total success of a COA element that we assume that their value controls the maximum value that p_T can achieve in the aggregation function. In effect, $s = 0$ between such risk drivers and all other risk drivers in the outer shell of this function; such an aggregation function is equivalent to a fixed production function in economics. Nested within this function, we then seek an aggregation function that determines the value of p_T that can be achieved by controlling other risk drivers. By definition, in this function, the value of p_T is limited so that $p_T \le p_{i'}$ for any risk driver i' that is critical to the total success of a COA element.

Through experimentation, we identified the following aggregation function:

$$p_T = \left[\sum_{i=1}^{n} r_i p_i \right]^{(1+\frac{k}{s})}$$

(Eq. 4.4)

in which there are n risk drivers that are not critical to the total success of the COA element, $\Sigma_i\, r_i = 1$ for all noncritical risk drivers, s is the single measure of substitutability among the n noncritical risk drivers, and k is an arbitrary positive constant.

For values of p_i and p_T shown in Equation 4.1, Equation 4.4 meets all the conditions shown in Equations 4.2 and 4.3 for each of the n noncritical risk drivers when all critical risk drivers are

---

marginal products—that is, holding output constant, the percentage change in the ratio of the inputs when the ratio of their marginal products changes by one percent.

[2] Hirofumi Uzawa, "Production Functions with Constant Elasticities of Substitution," *Review of Economic Studies*, Vol. 29, No. 4, October 1962, pp. 291–299, and Daniel McFadden, "Constant Elasticity of Substitution Production Functions," *Review of Economic Studies*, Vol. 30, No. 2, June 1963, pp. 73–83. In our notation, the constant elasticity of substitution function looks like this:

$$p_T = \left[\sum_{i=1}^{n} r_i p_i^{(\sigma-1)/\sigma} \right]^{\sigma/(\sigma-1)}$$

σ is the elasticity of substitution, which can vary in value from zero to infinity. When $\sigma = 0$, no substitution can occur among inputs in the production of any output. There is only one way to aggregate inputs into an output. When σ goes to infinity, inputs are all in effect identical to one another—"perfect substitutes." They can be used in any combination to yield an output.

fully under control (or for a value of p_T up to the minimum value of $p_{i'}$ for any critical risk driver i'):

$$\frac{\partial p_T}{\partial p_i}\frac{p_i}{p_T} = \left[1 + \frac{k}{s}\right]\left[\frac{r_i p_i}{\left[\sum_{i=1}^{n} r_i p_i\right]}\right] > 0$$

(Eq. 4.5)

$$\frac{\partial p_T}{\partial r_i}\frac{r_i}{p_T} = \left[1 + \frac{k}{s}\right]\left[\frac{r_i p_i - r_j p_j}{\left[\sum_{i=1}^{n} r_i p_i\right]}\right] > 0 \text{ if } r_i p_i - r_j p_j > 0$$

(Eq. 4.6)

$$\frac{\partial p_T}{\partial s}\frac{s}{p_T} = -\frac{k}{s}\,ln\left[\sum_{i=1}^{n} r_i p_i\right] > 0$$

(Eq. 4.7)

In Equation 4.6, we increase the importance of the ith risk driver solely at the expense of the jth risk driver. For any given set of values of p_i, $r_i p_i/(\Sigma_i\, r_i p_i)$ is the relative contribution of control of the ith risk factor to the total degree of control achieved by controlling all noncritical risk drivers together.[3] It is in effect proportional to the government's ability to manage problems associated with the ith risk driver.

Equation 4.7 holds because $\Sigma_i\, r_i\, p_i < 1$ and so $ln\,(\Sigma_i\, r_i\, p_i) < 0$.

For any fixed values of r_i and $p_i < 1$, the value of p_T rises from 0 when $s = 0$ to $\Sigma_i\, r_i p_i$ as s approaches infinity.

Equations 4.4 to 4.7 highlight the role of k as a scaling factor. They suggest that unity is an intuitively satisfying value of k. Choosing this value effectively scales s relative to unity, the value of the standard elasticity of substitution, σ, that economists use as the most common reference point with regard to substitutability. When $\sigma > 1$, economists speak of unusually high substitutability and conversely for $\sigma < 1$. $\sigma = 1$ for the classic Cobb-Douglas function broadly used in economics to aggregate inputs into outputs. With this in mind, we adjust Equation 4.4:

$$p_T = \left[\sum_{i=1}^{n} r_i p_i\right]^{(1+\frac{1}{s})}$$

(Eq. 4.8)

---

[3] In our notation, the analogous expression for the constant elasticity of substitution function is
$$r_i p_i^{(1-\frac{1}{\sigma})}\Big/\left[\sum_{i=1}^{n} r_i p_i^{(1-\frac{1}{\sigma})}\right].$$

When we apply Equation 4.8, we must keep in mind that s is not strictly speaking a direct measure of the standard economic elasticity of substitution, σ, but rather a measure of substitutability among noncritical risk drivers that is scaled similarly to σ. So, it is intuitively satisfying to note that, in fixed proportions, where σ would equal zero, $s = 0$. And, when two risk drivers are perfect substitutes for one another, where σ approaches infinity, s approaches infinity. As a practical matter, though, in our setting, it is best to use the graphical representation in Figure 4.1 to represent how alternative values of s affect the behavior of the aggregation function in Equation 4.8.

Figure 4.1. Relationship Between p_T and $\Sigma_i\, r_i\, p_i$ for Different Values of s

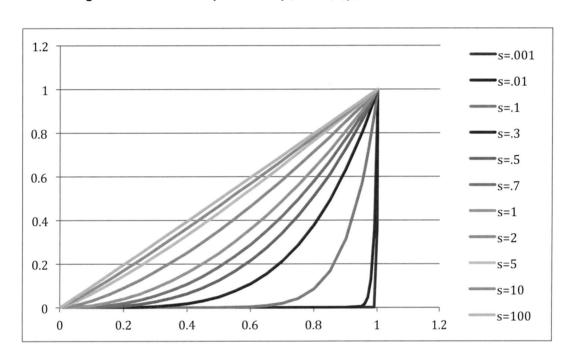

5. Formal Elicitation of Information on Risk Associated with COA Elements

This chapter explains how to elicit detailed professional judgment from relevant SMEs for each risk driver relevant to a COA element. This judgment covers the relationships among these risk drivers and the likelihood that the players relevant to the success of each COA element would be able to mitigate or even overcome the risks presented by each driver. In particular, this chapter describes the professional judgments to be collected and methods available to collect these judgments. It includes materials used to conduct the elicitations and the workshop procedure.

Information Elicited

For each COA element, our approach seeks professional judgments on the following attributes of each risk driver, which we initially identified in Chapter Three.

- **Relevance:** Does the risk driver have any influence on the likelihood that the government can successfully implement the COA element? Answer: yes or no.
- **Likelihood:** If the risk driver is relevant, what is the likelihood that the government will never fully mitigate the effects of this risk driver on the performance and cost of the program supported? If the government can fully mitigate these effects, what is the likelihood that it will do so within two years, between two and five years, or in more than five years? Answers: percentages that, taken together, must sum to 100 percent.
- **Partial substitutability/noncriticality:** If the risk driver is relevant, and the government cannot fully mitigate the effects of the risk driver, can mitigation of the effects of other risk drivers enable the government to implement the COA element in a way that ensures an acceptable level of system performance? That is, can the government substitute mitigation of the effects of other risk drivers for a failure to mitigate the effect of this one? Answer: yes or no.
- **Importance:** If the government can substitute mitigation of the effects of other risk drivers for a failure to mitigate the effects of this one, how important is this risk driver to successful implementation of the COA element relative to the importance of other risk drivers whose effects need not be fully mitigated for the government to successfully implement the COA element? Answer: ranking of importance, with ties allowed.

Answers to these questions allow us to assess how much of the full potential of a COA element the government can realize at any point in time.

Methods Used to Elicit Information

Our approach uses a series of carefully structured workshops to elicit this information from government representatives, contractor representatives, and an independent panel of SMEs chosen by RAND. In the F-22 BCA, the SMEs that RAND chose were knowledgeable about F-22 sustainment, but not directly associated with ongoing government or contractor activities relevant to the F-22 program. The workshops elicit information from each of these groups separately to help ensure that each group can freely provide expert information, free of any institutional bias. Additionally, the elicitation worksheets that are used to collect the SMEs' responses are kept anonymous. We organize all of these workshops in the same way.

Before we hold a risk workshop, we send the SMEs "homework" materials consisting of

- *read-ahead materials* containing background information about the goals of the study, definitions of the relevant COAs, and definitions of the risk drivers
- *instructions* on how to fill out a set of risk worksheets that prompt our subjects to provide the risk assessment inputs
- the *risk worksheets* themselves
- an *expertise questionnaire* asking about SMEs' relevant expertise
- for SMEs filling out worksheets on COA elements relevant to the air vehicle, an *appendix* with additional information about the risk drivers and COAs.

Appendix A presents an example packet including all of these materials. We pilot-test these materials to ensure SMEs understand them and can use them as intended.

Risk worksheets elicit the professional judgments, for each COA element, on the four attributes of a risk driver listed above. We asked SMEs to consider the risk drivers in relation to their ability to influence a successful implementation of the COA element, defined as an implementation that would sustain the primary performance goal relevant to the BCA (in the F-22 BCA, this ensured that fleet operational availability remained at the CY 2013 level). We asked SMEs to note that, in this definition, successful implementation of the COA element might increase the cost of product support or worsen performance metrics other than those relevant to the primary goal. Furthermore, we asked SMEs to consider the ability of risk drivers to reach *sufficiency,* or the level of mitigation of the effects of a specific risk driver needed to ensure product support of the primary performance goal relevant to the BCA.

Even though our tailored expert models rarely suggested that all eight risk drivers listed in Chapter Three were relevant to any particular COA element in the F-22 BCA, we chose not to give the participants in our new round of risk workshops too much guidance on what risk drivers to consider with regard to any particular COA element. To avoid anchoring their responses where possible, we offered the full list of risk drivers from Chapter Three to consider with regard to each COA element and then gave them the option of adding additional risk drivers to the worksheet if they chose to. A number did, but the list from Chapter Three captured the risk drivers relevant to most of our new SMEs.

Figure 5.1 presents a notional example of a completed risk worksheet. SMEs complete and return these initial worksheets in time for their inputs to be summarized and tabulated for the group workshop. Group workshops take place approximately one week after SMEs return their risk worksheets. We follow a structured protocol for the risk workshops, as shown in Appendix B. First, we provide a review of the *read-ahead materials* and *instructions* to fill out the risk worksheets. We also give SMEs a copy of the *risk worksheets* they had filled out previously. SMEs have the opportunity at this time to ask clarifying questions and ensure that everyone in the group has a common understanding of the COA element and risk driver definitions, as well as the risk worksheet instructions. For example, we probe SMEs by asking, "Was there anything confusing or that needs to be added or deleted [to the COA element or risk driver definition]?"

Figure 5.1. Notional Example of a Completed Risk Worksheet

Risk Driver	Relevant?		If relevant, is it partially substitutable?		If partially substitutable, Rank	Likelihood of Never Reaching Sufficiency	Likelihood of Reaching Sufficiency in...			SUM of likelihoods
							Less than 2 years	2-5 years	Greater than 5 years	
Attract and retain personnel	☒ Yes	☐ No	☒ Yes	☐ No	2	50%	50%	0%	0%	**100%**
Access technical data	☐ Yes	☒ No								**100%**
Access to IT and proprietary tools	☒ Yes	☐ No	☐ Yes	☒ No		10%	30%	30%	30%	**100%**
Access to IS systems	☒ Yes	☐ No	☒ Yes	☐ No	2	0%	50%	50%	0%	**100%**
Develop/ Institute new contracting vehicle	☒ Yes	☐ No	☒ Yes	☐ No	3	25%	50%	25%	0%	**100%**
Management of institutional knowledge	☒ Yes	☐ No	☒ Yes	☐ No	1	10%	10%	50%	30%	**100%**
Knowledge of/ relationship with the vendor base	☐ Yes	☒ No								**100%**
Adequacy of comparable sustainment processes	☐ Yes	☒ No								**100%**
	☐ Yes	☐ No	☐ Yes	☐ No						**100%**
	☐ Yes	☐ No	☐ Yes	☐ No						**100%**

NOTE: IS = information systems.

Next, we present anonymous results of the preworkshop risk worksheets. For each risk driver, we show the number of SMEs declaring it to be relevant or irrelevant and to be partially substitutable or not partially substitutable. We present the average, minimum, and maximum rankings of the risk drivers and then the normalized averages of the four likelihood values (such that they summed to 100 percent). In addition to summarizing SME inputs, we also provide an initial translation of these inputs into a scheduled risk assessment. We present the time to initial benefits accrual and time to full benefits accrual. Initial benefits accrual occurs when the government has mitigated problems associated with all critical risk drivers relevant to a COA element. In the absence of any critical risk drivers, initial benefits accrual occurs when the government has mitigated problems associated with any substitutable risk driver relevant to the COA element. Full benefits accrual occurs when the government has mitigated all problems associated with all risk drivers relevant to the COA element.

After presenting the results for one COA element, we lead SMEs through a group discussion. The discussion is meant to allow SMEs the opportunity to present their rationales for their chosen inputs to the group, as well as to learn about other SME perspectives. Therefore, we probe SMEs with questions such as, "The average likelihood here [for the risk driver never reaching sufficiency] was X. Could someone please tell me why they thought the likelihood of this risk driver never being able to reach sufficiency was less than X? More than X? Can someone tell me about a scenario in which this risk driver would never reach sufficiency?" The discussion also allows us to challenge extreme SME views in ways designed to widen the SMEs' perspectives and perhaps help them think beyond their initial biases. For instance, we ask SMEs who overwhelmingly state that a certain risk driver would never reach sufficiency, "Is there any scenario in which this risk driver could reach sufficiency? How likely is that scenario to occur?"

Once a workshop group has discussed all relevant COA elements, we provide SMEs with new risk worksheets. We instruct SMEs to think about all they have learned over the course of the workshop and complete the blank risk worksheets. We tell them that they can completely or somewhat revise their answers from their initial worksheets or copy their original answers exactly. Upon computing how these new inputs translate into the risk assessment results, we provide these to SMEs for their review and offer them the opportunity to have a follow-up discussion.

Table 5.1 presents summary information on a notional elicitation of professional judgment from four SMEs' knowledge about one COA element. In the first part of the assessment, we ask the four participants to make a determination on the relevance of each risk driver. All four participants deem personnel, data, IT, contracting, relationships, and the sustainment processes to be relevant to this COE element. Three of the four participants do not see relevance in information systems risk; one participant does not see relevance for knowledge risk. The determination of relevance to a large extent shapes or tailors the risk drivers to the model. In the next part of the assessment, we ask the four participants whether the risk driver is substitutable or not. Most of the participants feel that personnel, data, IT, and information systems could not be substituted. For the other risk drivers, there was a sense of substitutability or "work-arounds" that could be implemented. Rank ordering of the importance of the risk drivers suggests a decreasing ordering of importance from personnel, knowledge, data, relationships, contracting, and sustainment product support processes.

The final part of the risk assessment identifies the time it would take to reach a level of sufficiency. More specifically, for those participants who say that a risk driver is relevant, the time in which a degree of sufficiency is achieved to mitigate problems relevant to the COA element is summarized in the rightmost four columns. For example, according to the participants' determination, there is a 49 percent chance that the government will never fully mitigate problems associated with the personnel risk driver. There is only a relatively small chance that this will happen before five years (10 percent chance), and if it happens, it will likely take more than five years (41 percent chance overall.)

Table 5.1. Example of SME Input on Risk Drivers for a COA Element

	Respondents Stating Driver Is Relevant/ Irrelevant	Respondents Stating Driver Is Substitutable/Not Substitutable*	Average Rank (Minimum/ Maximum)	Expected Years to Implementation			
				Never	Less Than Two	Two to Five	More Than Five
Personnel	4 / 0	1 / 3	1 (1 / 1)	49%	4%	6%	41%
Data	4 / 0	1 / 3	2 (2 / 2)	26%	11%	31%	32%
IT	4 / 0	0 / 4		69%	1%	12%	18%
Information Systems	1 / 3	0 / 1		80%	5%	5%	10%
Contracting	4 / 0	3 / 1	2.3 (1 / 3)	15%	19%	48%	18%
Knowledge	3 / 1	3 / 0	1.7 (1 / 3)	18%	17%	25%	40%
Relationships	4 / 0	3 / 1	2 (2 / 2)	14%	17%	31%	38%
Sustainment	4 / 0	4 / 0	3.8 (3 / 4)	17%	11%	32%	40%

* This column only includes respondents stating that the driver is also relevant.

The information in the table can be combined to build a representation of the relevant risk drivers and a formulation of the probability over time that benefits accrue from a COA element. For example, Figure 5.2 shows the combined information in graphic form on the SMEs' assessment that the government will accrue benefits from a hypothetical COA element. This benefits accrual is identified in two ways. The time for initial benefits accrual, as defined above, is shown on the left side of the figure. The time for full benefits accrual, as defined above, is shown on the right. Reviewing the figure, the respondents generally judge that the government will probably (1) begin to accrue benefits from the COA element only after two years and (2) never accrue the full benefits potentially available from the COA element.

Caveats Relevant to Using Input from SMEs

Even when structured by a well-developed model, many common and recurring problems show up in risk assessments. Many result from the elicitation process used to obtain subjective assessments. These recurring problems include, among others, overconfidence, inaccuracy, aggregation, and interpretation problems.[1] *Overconfidence* occurs when SMEs overestimate the

---

[1] See Robyn M. Dawes, *Rational Choice in an Uncertain World*, New York: Harcourt Brace Jovanovich, 1988.

quality of their assessments. When this occurs, there is not enough uncertainty expressed in their assessments. *Inaccuracy* occurs when improper framing allows risk assessments to vary within an individual (internal inconsistency) and across individuals who might otherwise see things the same way (external inconsistency). *Aggregation* is a common problem that occurs when different assessments across risk areas are combined (sometimes averaged) in ways that lead to misleading, sometimes highly misleading, end results. Again, an understatement of the presence of uncertainty can compromise useful findings. In many cases, it is better to leave risk assessments disaggregated and present the spectrum of results. *Interpretation* is also often very difficult—once a risk assessment is made, what does it really mean, and how can it be used to help shape decisions that need to be made?[2] Clearly the results suggest that there was a high degree of uncertainty internal to many SMEs as well as across the SMEs. This should not come as a large surprise, since the assessments provided are subjective, where the historical data are largely inconclusive. Ways to mitigate the risk usually exist, but the question then becomes: "Is the mitigation worth it or should we live with the risk?" To resolve this question requires some knowledge of the "utility function" of the decisionmaker.[3]

Figure 5.2. Example of SME Assessment of Successful Implementation

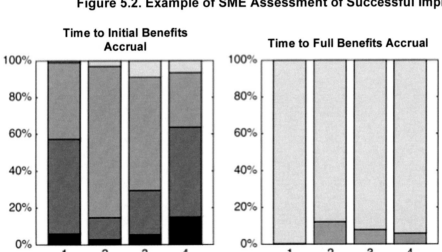

[2] See Cameron A. MacKenzie, "Summarizing Risk Using Risk Measures and Indices," *Risk Analysis*, Vol. 34, No. 12, 2014, pp. 2143–2162; Louis Anthony Cox, Jr., "What's Wrong with Risk Matrices?" *Risk Analysis*, Vol. 28, No. 2, April 2008, pp. 497–512; and Louis Anthony Cox, Jr., Djangir Babayev, and William Huber, "Some Limitations of Qualitative Risk Rating Systems," *Risk Analysis*, Vol. 25, No. 3, June 2005, pp. 651–662.

[3] See Robert T. Clemen and Terence Reilly, *Making Hard Decisions*, Cengage Learning, 3rd edition, 2014, and Thomas S. Wallsten and David V. Budescu, "State of the Art—Encoding Subjective Probabilities: A Psychological and Psychometric Review," *Management Science*, Vol. 29, No. 2, February 1983, pp. 151–173.

6. Integrating the Pieces to Characterize a COA

The elements of analysis described above give us everything we need to integrate information available on the benefits, costs, and risks associated with a COA element into a summary image that senior leaders can use to compare COA elements. This chapter describes the mechanics of bringing all these moving parts together. It first explains the series of steps the approach uses to execute a Monte Carlo analysis of each COA element. It then discusses how the approach brings information about the individual elements of a COA together to view the COA as a whole.

Steps to Integrate Information About One COA Element

Our risk assessment of any one COA element is informed by data on potential costs and savings associated with the COA and the professional judgments of SMEs about how likely it is that the government will implement the COA element to realize its full potential net benefits. For any set of SMEs, the following algorithmic set of steps explains how our approach generates findings that senior leaders can use on each of the COA elements examined in a product support BCA. Figure 6.1 summarizes the steps below.

1. Choose a COA element.
2. Use information elicited from the relevant set of SMEs to identify relevant risk drivers, which drivers are substitutable, and the relative importance of substitutable risk drivers (Chapter Five).
3. Use information from cost accounts to identify relevant costs and savings and conditions under which they apply for each risk driver in each year over the period of analysis (Chapter Two).
4. Use the following steps to generate one future for this COA element.
5. Determine the value of p_{Tt} for each year during the period of analysis.[1] To do this, perform Steps 6 through 10 for each year.[2]
6. Choose a risk driver (Chapter Three).
7. Make a random draw to determine whether the Air Force ever fully mitigates problems associated with this risk driver and, if so, when during the period of analysis this occurs, using information elicited from relevant SMEs to define the probability distribution used in the random draw. To do this, determine whether mitigation occurs within a segment of zero to two years, two to five years, or beyond five years. Within the segment during which mitigation occurs, make a second random draw, from a uniform distribution across

---

[1] These steps add a time index, t, to p_i and p_T from Equation 4.8, to emphasize that this approach determines values of these for each year during the period of analysis.

[2] Step 5 does not appear in the figure, because it essentially comprises Steps 6 through 10, which do appear in the figure.

the segment, to determine when precisely during the segment mitigation occurs. Use this draw to set the value of p_{it} for each year in the period of analysis. $p_{it} = 0$ for each year up to the one in which all problems are mitigated. After then, $p_{it} = 1$ (Chapter Five).

8. Drawing on Steps 3 and 7, use business rules to identify costs and savings relevant to the risk driver in each year during the period of analysis. For example, in the F-22 BCA, investments in technical data, information systems, and design of performance-based logistics (PBL) agreements occur in year one. Training costs occur in each year for which $p_{it} = 0$ for new government employees. Annual savings from substituting government employees for contractor employees start in the first year for which $p_{it} = 1$ for government employees. Annual savings from insourcing material management begin as soon as $p_{it} = 1$ for any risk factor relevant to material management that is not considered substitutable—for example, access to technical data. Annual savings from implementation of a PBL or improved contractor processes begin in the first year in which $p_{it} = 1$ for such changes.

9. Return to Step 6, choose another risk driver, repeat until all risk drivers relevant to the COA element in Step 1 are addressed.

10. Calculate the value for p_{Tt} for each year in the period of analysis, starting with year one. $p_{Tt} = 0$ in any year in which $p_{it} = 0$ for any risk factor that is not substitutable. Moving through time, once $p_{it} = 1$ for all risk factors that are not substitutable, apply information from Steps 2 and 7 to Equation 4.8 to calculate the value of p_{Tt} for each remaining year during the period of analysis (Chapter Four).

11. Working through the period of analysis, multiply the cost or savings in each year from Step 8 by the value of p_{Tt} for that year from Step 10 to generate a sequence of realized cash flows for one future (Chapter Two).

12. Use the appropriate, risk-free OMB discount rate to calculate the NPV of this sequence. This is the NPV for one future (Chapter Two).

13. Return to Step 4, generate another future, and repeat until all futures desired are generated.[3]

14. Use the NPVs generated by this process for different futures to create a subjective probability frequency distribution for the COA element (Chapter Two).

15. Use the distribution in Step 14 to generate policy-relevant summary statistics for NPV, such subjective confidence intervals, expected value, and probability that NPV is negative (Chapter Two).

16. Return to Step 1, choose another COA element, and repeat until all COA elements are addressed.

---

[3] Additional futures should be generated until the subjective probability frequency distribution being constructed stabilizes. For example, when we constructed distributions for individual COA elements, we found that repeated draws of only 100 futures yielded significantly different distributions. We did not find stability until we built each distribution based on 500 draws. Given the low cost of adding each future, we decided to build each distribution with draws for 1,000 futures to build in a margin of safety. Formal methods are available to choose an optimal number of draws. See, for example, George S. Fishman, *Monte Carlo: Concepts, Algorithms, and Applications*, New York: Springer-Verlag, 1996; Averill M. Law and W. David Kelton, *Simulation Modeling and Analysis*, 3rd ed., New York: McGraw Hill, 2000; and Mustafa Y. Ata, "A Convergence Criterion for the Monte Carlo Estimates," *Simulation Modelling Practice and Theory*, Vol. 15, No. 3, March 2007, pp. 237–246. But these methods require a subjective judgment of how much stability in a subjective distribution is acceptable. Given the low cost of additional computations, we simply chose a conservative number likely to yield enough stability to give decisionmakers confidence in the distributions generated.

Figure 6.1. Integration of Information on One COA Element

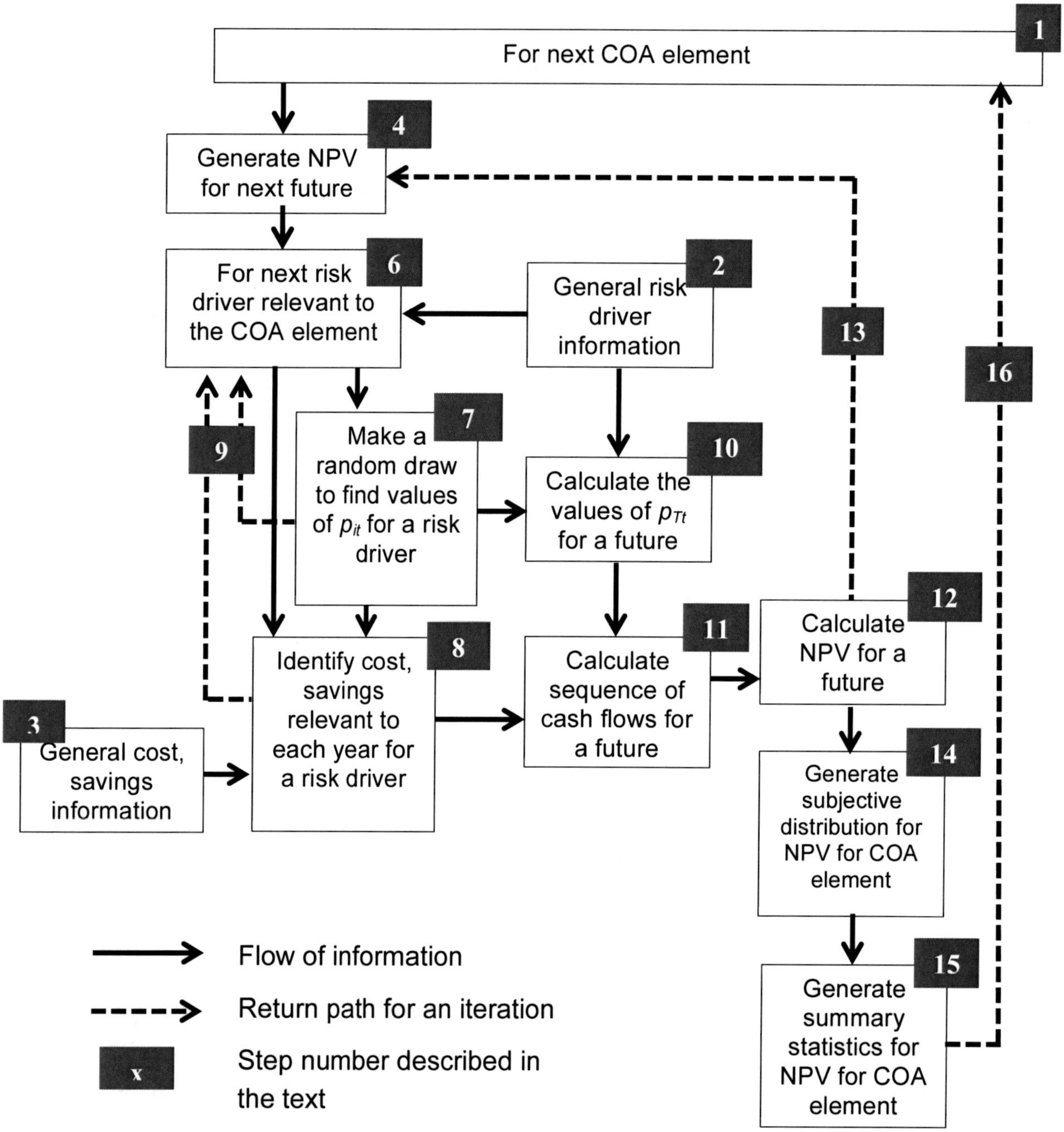

Figure 6.2 displays an example of the kind of information that this algorithm can produce for one COA element. The top panel displays 90-percent subjective confidence intervals and expected values for 12 individual SMEs—five with no formal current affiliation with the program under review, four employed by the contractor currently providing support for the program, and three employed in government activities associated with the program. The red line corresponds to the lowest value of NPV possible. The black horizontal line at the top

corresponds to the NPV for an implementation in which everything goes right. The bottom panel displays consensus independent, contractor, and government confidence intervals and expected values derived from the intervals in the top half.

Several things are worth noting. First, great uncertainty appears to exist about the value that the government might derive from this COA element. The views of independent SMEs are especially diffuse. Second, despite this uncertainty, discernable differences appear to exist across the three groups of SMEs. Those affiliated with the contractor appear to be far more skeptical than those associated with the government. Third, differences of opinion also appear to exist within each group; however, there appears to be enough consensus within each group to derive a group confidence interval to represent each group.

Our experience teaches us that this sort of outcome is likely to occur. We conclude that all points of view should be assessed. When comparing COA elements, it is worth asking whether independent, contractor, and government SMEs *rank* the competing COA elements differently. Even in the face of uncertainty and disagreement like that shown in Figure 6.2, we have found much greater agreement—robustness—on the ranking that such groups give competing COA elements.

Figure 6.2. Illustrative 90-Percent Confidence Intervals for One COA Element

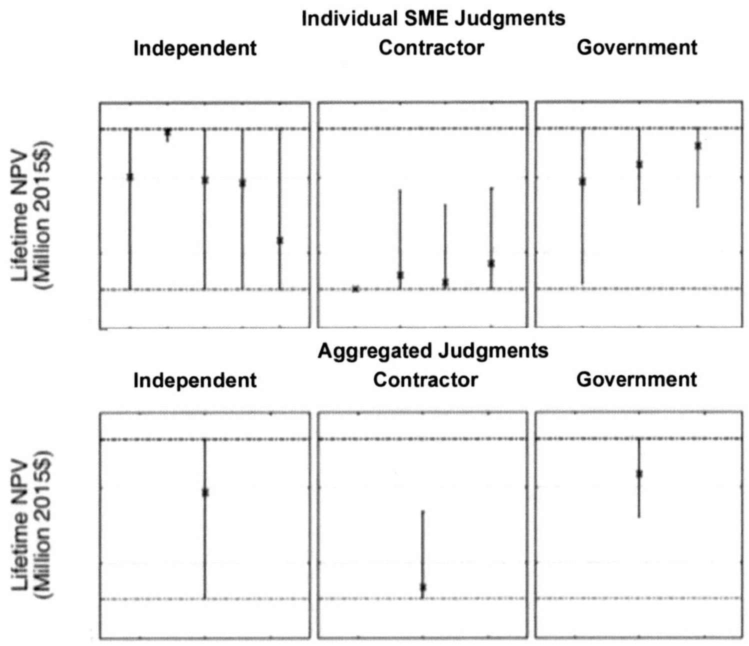

Integrating Information Across Elements Within a COA

Imagine a COA with four elements. For example, one insources a small analytic task. The second insources materiel management for a major subassembly, such as a landing gear. The third offers a set of contractor process improvements in return for a firm fixed price requirements agreement over five years on a selected set of tasks. A fourth offers a PBL agreement on a major subassembly, such as a fire-control radar. The approach described here offers a way to develop a subjective probability frequency distribution like that in Figure 2.4 for each COA element. The approach could generate confidence intervals and expected values, such as those shown for Figure 6.2, for each COA element. Ultimately, however, DoD BCA guidance asks for information about an entire COA.[4] How can we move from information about each of the elements within a COA to an assessment of the COA as a whole?

This is a challenge for at least two reasons. First, it is likely that the SMEs who understand one of these COA elements well do not understand the others as well. An effective assessment of all these COA elements is likely to draw on different SMEs to conduct studies of different COA elements. That means that it will be impossible to build confidence intervals like those in the upper half of Figure 6.2 for COAs with disparate elements in them.

Even if the same SMEs assessed the risk drivers associated with each COA element, we would have difficulty defining, in practical terms, what any single future means when we look across all COA elements at the same time. The COA elements would likely involve different risk drivers and mitigations in different ways. For example, the dominant risk driver for the insourcing would likely involve personnel management, while the dominant risk driver for the fire-control-radar PBL agreement might concern the government's ability to obey the spirit of the agreement over a five-year period. Even if the risk drivers were similar, personnel management could generate different problems in an activity that the government understands well and in another that it does not. Such subtleties lie at the heart of the tailored expert models described in Chapter Three.

A more extensive analysis of risk drivers might identify exogenous shocks that would affect different COA elements in similar ways, thereby inducing correlations in the subjective probability distributions for any two COA elements in the same COA. Standard Monte Carlo methods could then be used to capture these correlations in the analysis. We have not attempted to do that. Rather, we implicitly assume that the subjective probability frequency distributions for any two COA elements are statistically independent. To create a distinct future for one COA as a whole, we randomly draw a value of NPV from the distribution for each element of the COA and sum the values to yield the value of NPV for the COA as a whole in that future. Taken together, many similar draws yield a distribution for the COA as a whole, which we use to represent the total COA in BCA analysis. We can preserve consensus, independent, contractor,

---

[4] "Comparison of Alternatives," Section 4.9.1, in Office of the Principal Deputy Assistant Secretary of Defense for Logistics and Materiel Readiness, 2011.

and government distributions for each COA, but not distributions for individual experts. There are important limitations to this approach that are worth noting. Most notably for this application, aggregation of SME judgments results in a loss of information that may be important to decisionmakers. Information such as agreement or disagreement among SMEs will not be reported as part of aggregated distributions. Furthermore, aggregated distributions that show large amounts of uncertainty may result from either SME agreement, with each SME providing distributions with wide uncertainty bands, or SME disagreement (with limited or wide uncertainty bands). To address this limitation, we report aggregated results with a qualifying explanation of whether the uncertainty shown is a result of disagreement or uncertainty in the individual distributions.

Figure 6.3 provides an example of assessments of the risk-adjusted savings associated with three COAs that we examined in the F-22 sustainment BCA. The information displayed here is a bit different from that shown in Figure 6.2. The blue boxes show a 50-percent confidence interval, the red lines show the median NPV, and the "x" marks the mean NPV for each distribution. The dotted horizontal line shows the maximum NPV available from each COA if implementation is flawless. As in Figure 6.2, we see high uncertainty and significant differences in the assessments of COAs across groups of SMEs. Our analysis allows us to explain such variation to help decisionmakers understand the primary drivers behind the judgments shown.

Figure 6.3. Risk-Adjusted Savings Assessment for Three Alternative COAs

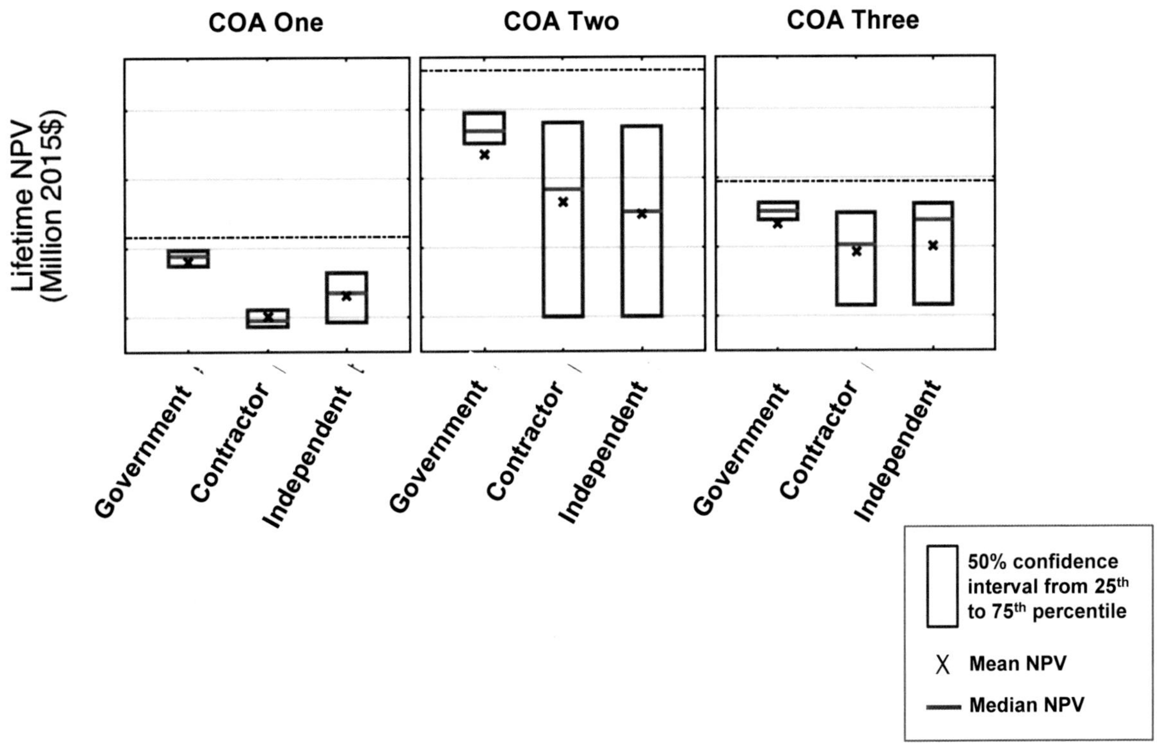

41

Figure 6.4 shows the same information as Figure 6.3, but removes one source of risk at a time to reveal the effect each source has on the distribution for each COA. For example, the BaseCase distribution for COA One is the same as the distribution for independent SMEs in Figure 6.3. The data distribution for COA One in Figure 6.4 removes data-related issues as sources of risk, the IT distribution for COA One removes IT-related issues, and so on. The results in Figure 6.4 indicate that none of these risk sources has a significant marginal effect on the distribution for COA One, suggesting that different assumptions about these risk sources would not have a large effect on the distribution presented to decisionmakers.

The same cannot be said of the sources of risk relevant to COAs Two and Three. Risks associated with contracting stand out in each case. If these COAs could be adjusted to reduce contracting risk, their downside risks would fall significantly. That does not necessarily mean that these COAs could be changed to mitigate contracting risk. In this case, for example, the independent SMEs were highly uncertain as to whether the government could design and execute an effective contract, even if it wanted to. Such information was helpful to decisionmakers.

Figure 6.4. Effects of Assumptions about Individual Sources of Risk

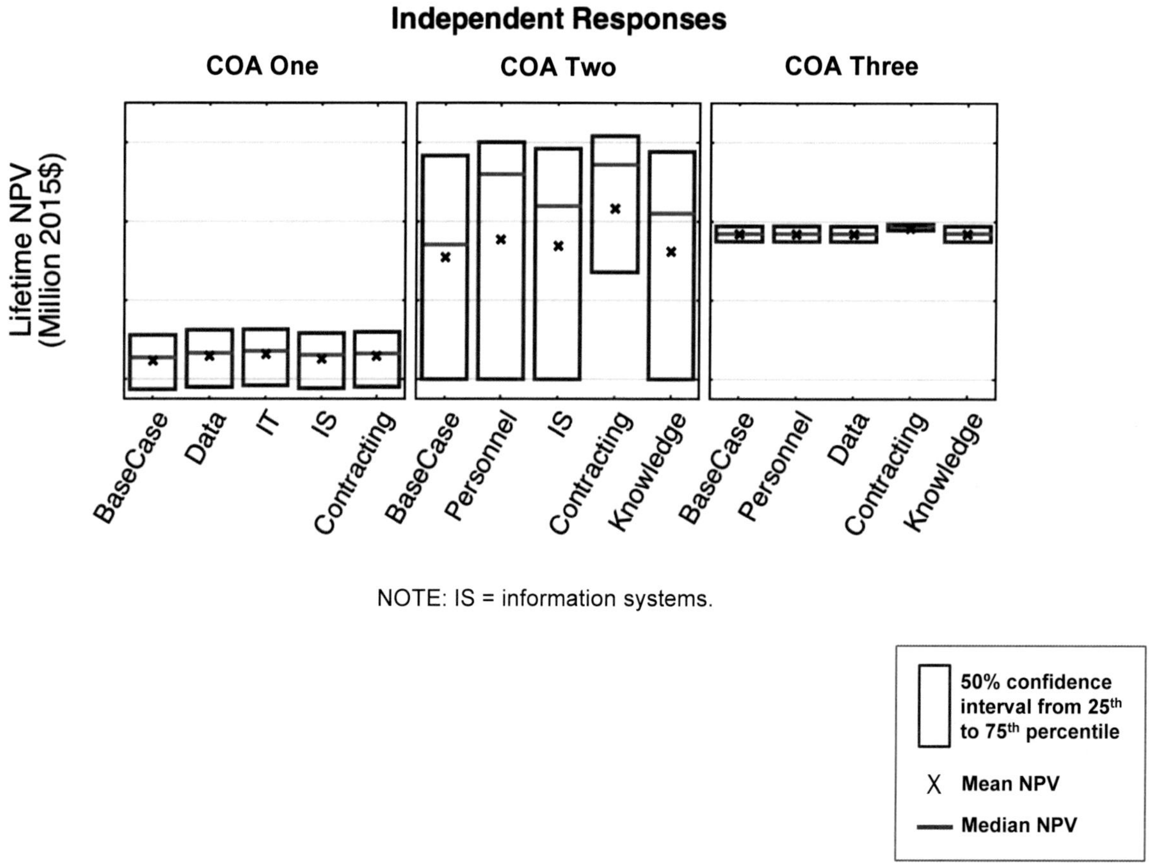

NOTE: IS = information systems.

7. Providing Support for Senior Decisionmakers

Our approach is designed to generate information that senior decisionmakers can use to choose among BCA COAs. It does not preempt their authority or responsibility to make these choices. But BCA guidance asks the analysts in a BCA to recommend a preferred COA and justify that recommendation.[1] To respond to that request, we provide guidance on how senior decisionmakers can best use the information that this approach generates. It considers two situations. In the first, decisionmakers agree to emphasize the primary attribute of performance described in Chapter Two and seek to use that attribute to choose among COAs. In the second, at least some decisionmakers believe a secondary attribute of performance deserves serious attention when choosing among COAs.

Choosing Among COAs When the Primary Attribute of Performance Dominates

Our approach generates subjective probability frequency distributions that senior decisionmakers can use to choose among COAs. These decisionmakers may find it useful to rely on summary statistics derived from these distributions. Figure 7.1 illustrates a case where decisionmakers rely on 90-percent subjective confidence intervals to compare three COAs. Only 5 percent of the outcomes identified in the risk assessment lie beyond either end of each of these intervals. The value of COA NPV over the period of analysis appears on the horizontal axis. The intervals appear as three bars. Recall from Chapter Two that subjective confidence intervals are of special interest for decisionmakers who are risk-averse and who care about where the general central tendency of NPV for a COA is likely to be.

Note that considerable uncertainty exists about each COA and that there is a significant chance of negative NPV. These circumstances are common in our experience.

It appears likely that the general central tendency for the NPV of COA Two lies above that for COA One. Therefore, COA Two is likely to dominate COA One. By a similar argument, COA Three is even more likely to dominate COA One. The relative position of NPV for COAs Two and Three is harder to discern. But COA Three presents a significant probability of a negative NPV. If decisionmakers worry about negative outcomes, this in itself could lead them to prefer COA Two to COA Three, even though COA Three offers the potential for much larger positive outcomes than COA Three. Similarly, COA Two appears to present far less uncertainty in general than COA Three. If decisionmakers are highly risk averse, this could lead them to

---

[1] "Recommendations," Section 4.10, in Office of the Principal Deputy Assistant Secretary of Defense for Logistics and Materiel Readiness, 2011.

prefer COA Two to COA Three. If the possibilities of a negative outcome or a high degree of uncertainty do not concern decisionmakers, they will need to look more closely at COAs Two and Three. More-detailed information on their relative expected values or even their relative median values could be helpful. In the end, the subjective confidence intervals in Figure 7.1 can help senior decisionmakers explore their priorities in the context of a real decision, which often helps them sharpen their understanding of those priorities. At the end of that exploration, the information presented here should have helped them formulate a clear justification for their choice.

Figure 7.1. Subjective Confidence Intervals for Three Notional COAs

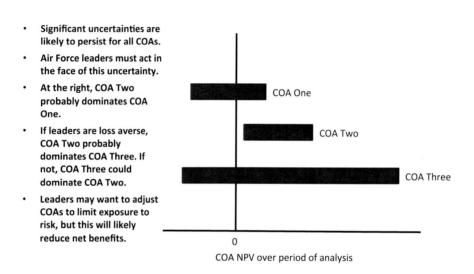

As analysts, we must remind decisionmakers that the best outcome is contingent on their priorities. We can offer recommendations that are contingent on alternative potential priorities. If the level of uncertainty that our approach reveals about alternative COAs is like that shown in Figure 7.1, which is what we would typically expect, our recommendations about these three COAs would look something like the discussion in the paragraph above.

Recommendations could also potentially look for further mitigations that decisionmakers could use to limit the negative outcomes associated with each COA. The time for seeking such mitigations is really during the risk analysis itself, which should seek to present decisionmakers with COAs where cost-effective mitigations have already been assessed and implemented as part of each relevant COA. As a practical matter, however, risk analysis typically proceeds iteratively and could continue to explore such mitigations as it presents interim findings to decisionmakers.

Choosing Among COAs When a Secondary Attribute of Performance May Be Important

In the second F-22 product support BCA, Air Force leaders agreed that the dominant attribute of performance for the BCA was the operational availability of the F-22 fleet. But other attributes of performance are also relevant. The harder it is to choose between two COAs based on the primary attribute of performance, the more likely it is that decisionmakers will want to explore how important they believe secondary priorities are.

For example, some senior Air Force leaders strongly prefer COAs that insource important elements of product support. Suppose that two COAs, one focused on continuing current contractor product support and the other focused on insourcing product support, present similar levels of NPV. Because the effects of the dominant attribute of performance have been fully monetized in the NPVs presented, the dominant attribute of performance is no longer relevant to the choice between COAs. Those who inherently favor insourcing would then have an opening to argue for their preferred COA. Of course, not all Air Force leaders agree that insourcing is inherently a good thing.

But even if the analysis described here indicated that insourcing would likely cost more than traditional contract product support, advocates of insourcing could use the approach described here to inform their discussion in the Air Force (or elsewhere in DoD). If the NPV of insourcing is N_I in any future and the NPV of traditional product support is N_T in that future, the difference in cost between the two in that future, $N_I - N_T$, is an amount that the Air Force would have to forego to favor the insourcing COA. If advocates of insourcing can make a compelling case that the inherent value of insourcing in a particular BCA is greater than $N_I - N_T$ in many futures, they can build a case for insourcing.

Note that the approach described here seeks to induce decisionmakers to express their priorities in the dollar terms that currently drive the programming and budgeting systems in DoD. Senior decisionmakers are familiar with stating their preferences in such dollar terms. That makes it easier for decisionmakers to present an argument or understand an argument presented in such terms, particularly for attributes of performance often discussed primarily in qualitative terms. This presents the findings of a BCA in terms that can directly inform ongoing decisionmaking.

Appendix A. Risk Workshop "Homework" Materials

This appendix provides the text of risk workshop "homework" materials used in the F-22 product support BCA. Details of the COAs considered in that BCA have been redacted to allow broad distribution of these materials. Minor changes have also been made in the text to make it consistent with the terminology used in the body of this report.

Read-Ahead Information for the F-22 Product Support BCA

This document will provide background information and definitions relevant to the risk workshop for the F-22 Product Support BCA. Please read the following information before filling out the risk worksheet.

Background

RAND is currently helping the Air Force conduct a product support BCA for the F-22. As a part of the BCA, a number of alternative COAs for sustainment of the air vehicle and F119 engine have been developed. The costs and benefits of these COAs will be compared for the period from 2018 through 2033. RAND has estimated the maximum potential savings for each COA (savings given that [its elements] are successfully [implemented] under the best case scenario). The goal of this risk workshop is to evaluate the extent to which those savings can be realized. RAND has identified a set of risk drivers that may influence the successful [implementation] of the [COA elements]. Each of these risk drivers may therefore impact the potential savings. As an F-22 sustainment SME, you will be assessing, for each COA [element], whether (1) these risk drivers carry influence on the successful [implementation], (2) how important they are to the successful [implementation] and (3) the likelihood associated with those risk drivers. The results of the risk workshop will be used to quantify the expected savings associated with each COA [element].

Courses of Action

RAND, the Air Force, and [contractor teams] have developed three air vehicle and three engine COAs for which you will be providing assessments. These six COAs and [their elements] are described below. [details on COAs redacted]

Risk Drivers

RAND identified eight risk drivers or factors that could influence the successful and timely [implementation] of [COA elements]. For each COA [element], you will be assessing on the risk worksheet the relevance, importance, and likelihood of occurrence of each risk driver. Some risk drivers are relevant to one COA [element], while other risk drivers are relevant to many. If you believe this list is omitting any important risk drivers, the worksheet allows you to add your own. If a majority of the group attending your risk workshop identifies the same risk driver, it will be added to the risk assessment and risk workshop discussion. The risk drivers are described below:

- **Attract and retain relevant personnel**: This driver involves the risk associated with the Air Force obtaining and maintaining personnel with the needed skills after transferring functions to the government. Given that the F-22 is a highly specialized platform, there may be personnel risk associated with the transfer of specialized functions. Subfactors include the ability to attract and retain qualified personnel, address salary differentials, and train personnel. An on-the-job training program with Lockheed Martin and Boeing personnel is envisioned.
- **Ability to access technical data**: This driver involves the risk associated with accessing technical data packages within each of the transferring functions, unless these technical data packages will be provided by the Lockheed Martin–Boeing team at no additional cost. The F-22 was designed and produced using a state-of-the-art paperless system. The technical data is housed within this proprietary system, and there is risk in transferring functions that is directly associated with accessing and/or transferring these data for Air Force use. This involves transferring both existing and future data. Since the F-22 was produced with a high level of concurrency, the number of modifications and changes to technical data are unusually high, relative to other aircraft in the Air Force inventory. [details of COA elements redacted]
- **Access to IT and proprietary tools**: This driver entails risk associated with gaining access to or being able to implement similar (or identical) IT and intellectually proprietary tools associated with the sustainment functions for the F-22. Specifically, the F-22 uses specialized tools to facilitate the execution of various functions. There is risk associated with the transfer of some functions, given that the Air Force does not have access to or currently operate these tools. The government needs to either have comparable tools and/or arrange access to Lockheed Martin and Boeing tools, for example, through a licensing agreement or other contract vehicle.
- **Access to information software systems**: This driver involves the risk of transferring information software systems from the Lockheed Martin–Boeing team to the Air Force. Specifically, the Lockheed Martin–Boeing team uses proprietary forecasting tools and tracks the percentage of the F-22 fleet that has each item or piece of equipment so it does not overpredict demand. Parts are ordered at the lowest cost for achieving aircraft availability for the F-22. The organic supply system processes do not do this. As you disaggregate supply chain management, you raise the possibility of buying more parts than needed.
- **Development of/instituting a new contracting vehicle**: This driver entails the risk associated with developing a long-term contract (five-year base with five-year option) that would be optimal for providing incentives to Lockheed Martin and Boeing to invest in cost-saving measures. It may be difficult for the Air Force to leave the terms and conditions of the contract intact for the duration of the contract.
- **Management of institutional knowledge of the F-22**: This risk driver is associated with Air Force competency and depth of knowledge in managing a specialized fifth-generation aircraft, particularly in addressing unanticipated problems. As with personnel, technical

data, and tools, given the specialization of the F-22, the Lockheed Martin–Boeing team has developed institutional knowledge in managing the F-22. While the lack of data and tools could be overcome by the Air Force, it may take more time to develop the base of knowledge to use the information in a productive and efficient manner. This includes where and who to go to in order to address unexpected problems; according to the Lockheed Martin–Boeing team, it takes years to develop the skills and knowledge to address these kinds of unanticipated problems.

- **Knowledge of and/or relationship with the vendor base**: This driver involves the risk associated with transferring different supply functions to the Air Force and Defense Logistics Agency. Specifically, Lockheed Martin and Boeing have established long-term agreements with vendors and have relationships with them that can allow for a robust supply base. They know who to go to, and when part of the supply base disappears (as it has in the past given the small platform count), they can find replacements relatively efficiently. In contrast, the government tends to provide a contract for each buy. It is a different kind of relationship, and eventually may be overcome, but in the meantime, the F-22 and the F-35 give the Lockheed Martin–Boeing team leverage in negotiations and relationships with vendors.
- **Adequacy of comparable sustainment processes**: This risk driver involves the ability of the Air Force to provide a sustainment process comparable to the one currently used by the Lockheed Martin–Boeing team. Similar to institutional knowledge, this focuses on the sustainment processes themselves. [detail on COAs redacted]

Instructions for the F-22 Air Vehicle and Engine Product Support BCA Risk Worksheets

This risk worksheet is intended to guide you through a structured process in which you will help to determine the likelihood of successfully [changing] F-22 sustainment functions for a set of COAs. Before answering the questions on the risk worksheet, please thoroughly review the read-ahead materials and these instructions. Please use your expert judgment based on past factual and relevant experience to determine your responses. No research or outside consultation should be necessary to complete this exercise.

You will be filling in a number of cells in a table that looks like the one on the next page. You will be filling in the same information for the following COA [elements]. Each COA [element] will be shown on a different page on the risk worksheet. You will fill out nine tables corresponding to the following COA [elements]. [details of COAs redacted]

The following instructions will explain how to fill out each column of [Table A.1].

Table A.1. Example of Risk Worksheet (Part 1)

Risk Drivers	Relevant?		If Relevant, Is It Partially Substitutable/ Noncritical?		If Partially Substitutable, Rank	Likelihood of Never Reaching Sufficiency	Likelihood of Reaching Sufficiency in. . .			Sum of Likelihoods
							Less Than Two Years	Two to Five Years	More Than Five Years	
	Yes	No	Yes	No						
Attract and retain personnel	☐	☐	☐	☐						100%
Access to technical data	☐	☐	☐	☐						100%
Access to IT and proprietary tools	☐	☐	☐	☐						100%
Access to information software systems	☐	☐	☐	☐						100%
Develop/ institute new contracting vehicle	☐	☐	☐	☐						100%
Management of institutional knowledge	☐	☐	☐	☐						100%
Knowledge of/ relationship with the vendor base	☐	☐	☐	☐						100%
Adequacy of comparable sustainment processes	☐	☐	☐	☐						100%
	☐	☐	☐	☐						100%
	☐	☐	☐	☐						100%

Column One: Risk Drivers

These are the risk drivers you reviewed in the read-ahead materials. Note that there are a few blank rows. If you believe a risk driver is missing from the risk assessment for one or more COA [elements], you may add it to the appropriate table(s) and fill out the new row(s) as you would the other rows. At the end of the worksheet, we ask that you provide an explanation for any additions you might make.

Columns Two and Three: Relevant and Partially Substitutable/Noncritical

For each risk driver, we would like to get a sense of (1) whether it is relevant for each COA [element] and (2) how important that risk driver is to the success of [implementation]. We define a **successful implementation** as one that would cause no harm to the fleet (i.e., result in equal availability to the existing structure). *Note that in this definition, a successful implementation may be one that increases the cost or worsens performance metrics other than availability—as long as availability remains the same as it is currently.*

- The second column asks whether the risk driver is **relevant** to that COA [element]. Here, you will assess whether the risk driver has any influence on the likelihood of a successful [implementation]. If there is no influence, check the box for not relevant. If there is even the slightest possibility of an influence, it should be deemed as relevant. Later stages of this worksheet will allow you to determine the extent of that influence. Please first make relevance determinations for ALL risk drivers BEFORE moving on to the next step. *If the risk driver is NOT relevant, you do not need to fill in any more cells in that row. You can put a strike through the entire rest of the row.*

- The third column asks whether the *relevant* risk drivers are **partially substitutable**, or **noncritical**. That is, whether lacking a full level of that risk driver would limit the Air Force's ability to successfully [implement] because *nothing can substitute completely* for that risk driver. For instance, could a [COA element] be successfully [implemented] if only a partial set of relevant personnel were attracted and retained or if only a partial transfer of data occurred? If nothing can substitute for these risk drivers, they would be considered NOT partially substitutable. Or is the successful [implementation] of a COA [element] possible with only a partial set of relevant personnel because the lack of personnel can be made up for with the proper IT tools? If that is the case, then personnel would be partially substitutable or noncritical. Please first make partially substitutable determinations for ALL risk drivers BEFORE moving on to the next step. *If the risk driver is NOT partially substitutable, you do not need to fill out the FOURTH cell in that row (ranking). You can put a strike through the fourth cell in that row. You DO need to fill out the likelihood cells in that row (fifth through eighth cells).*

Column Four: Rank

For the risk drivers that are partially substitutable, the fourth column asks you to rank the importance of each substitutable risk driver to successful [implementation]. A rank of one is the most important, two is the second-most important, and so on. If two risk drivers are equally important, you may rank them equally (i.e., as a tie). Only rank those risk drivers that are substitutable.

Columns Five–Eight: Likelihood of Successful Implementation

For the likelihood columns, we define the term *sufficiency,* or the lowest level of a specific risk driver needed to ensure a successful implementation with equal availability. For each *relevant* risk driver, we would like to explore the likelihood the Air Force (1) will *never* achieve sufficiency and (2) will achieve sufficiency within two years, between two and five years, or in more than five years.

- For the relevant risk drivers, the fifth column asks for the likelihood of *never* being able to achieve *sufficiency* of the risk driver. For instance, how likely is it that the data will *never* be able to achieve sufficiency under a COA [element]? When determining this likelihood, think about all of the scenarios that could lead to the data not being successfully [accessed by the government] and the likelihood of those scenarios. Remember, *successful* scenarios can include those that increase costs or reduce performance metrics other than availability.
- For the relevant risk drivers, the sixth through eighth columns ask for the likelihood that the Air Force will achieve *sufficiency* of the risk driver for that COA [element] in (1) less than two years, (2) between two and five years, or (3) more than five years. When determining these likelihoods, think about the least amount of time needed to achieve sufficiency of that risk driver (i.e., a sufficient level such that successful [implementation] will be highly likely). Also, think about the scenarios where it might take longer and how long that could take.

There will be four likelihoods for each risk driver. Since they represent every possible state of the world (i.e., never [implement], [implement] in less than two years, between two and five years, and more than five years), the four likelihoods will add up to 100 percent, as shown in the ninth column. Please first determine the likelihood that [the Air Force will not achieve *sufficiency* of the risk driver for that COA element], followed by the other three likelihoods of [successful sufficiency] within a specific time period.

Notional Example

[Table A.2] provides a notional example where three of the risk drivers are not relevant, one of the remaining risk drivers is not partially substitutable and two of the remaining risk drivers were thought to be of equal importance (and therefore received equal ranks). As shown below, we ask that you put a strike through the rest of the row when you determine a risk driver to not be relevant and that you put a strike through the fourth (Rank) column when the risk driver is not partially substitutable. You do not need to fill in those cells. This table also shows the two blank rows, suggesting that for this COA [element], no risk drivers were missing [detailed worksheets for COAs redacted].

Thank you for filling out this risk worksheet. If you added any risk drivers to the tables in the worksheet, please provide a description of those risk drivers below and why you believe they are relevant and not covered by the original eight risk drivers.

Table A.2. Example of Risk Worksheet (Part 2)

Risk Drivers	Relevant?	If Relevant, Is It Partially Substitutable/ Noncritical?	If Partially Substitutable, Rank	Likelihood of Never Reaching Sufficiency	Likelihood of Reaching Sufficiency in . . .			Sum of Likelihoods
					Less Than Two Years	Two to Five Years	More Than Five Years	
Attract and retain personnel	☒ Yes ☐ No	☒ Yes ☐ No	2	50%	50%	0%	0%	100%
Access to technical data	☐ Yes ☒ No	☐ Yes ☐ No						100%
Access to IT and proprietary tools	☒ Yes ☐ No	☒ Yes ☐ No	2	10%	30%	30%	30%	100%
Access to information software systems	☒ Yes ☐ No	☒ Yes ☐ No		0%	50%	50%	0%	100%
Develop/institute new contracting vehicle	☒ Yes ☐ No	☒ Yes ☐ No		25%	50%	25%	0%	100%
Management of institutional knowledge	☒ Yes ☐ No	☒ Yes ☐ No	1			50%	30%	100%
Knowledge of/ relationship with the vendor base	☐ Yes ☒ No	☐ Yes ☐ No						100%
Adequacy of comparable sustainment processes	☐ Yes ☒ No	☐ Yes ☐ No						100%
	☐ Yes ☐ No	☐ Yes ☐ No						100%
	☐ Yes ☐ No	☐ Yes ☐ No						100%

52

Expertise Questionnaire

Please answer the following questions about yourself. Note that we will only use these answers in summary statistics without any identifying characteristics. Any published reports will present summary statistics and nonattributional information.

Title at your current position _____

Years at your current position _____

How many years have you been involved with [individual activities relevant to COA elements]?

Please briefly describe your involvement in these [activities] below:

Appendix B. Risk Workshop Facilitation Protocol

F-22 Product Support BCA Risk Workshop Protocol

Introduction

Thank you for attending the F-22 Product Support BCA Risk Workshop for [COA element]. Here is our agenda for the day [*show slide*]. We'll first review the COAs and risk drivers and then discuss their definitions as a group, as well as discuss whether there are any risk drivers that were missing from the initial list. Then we'll review all of the fields in the risk worksheet. At that point, we'll show you the summary results from your initial risk worksheets. We'll discuss how the results differ between the group and why. Finally, I'll ask you to consider all that you heard today and fill out the worksheets again. *[If needed]* Note that the final results that we will report from this workshop will be a summary of the individual postworkshop worksheets you fill out. We will not count your initial assessment.

Explanation of the COAs

Now we will present the COAs we'll be considering today. There will be a chance in a few minutes to discuss these COAs, so please hold comments and questions.

[*review slides*]

Explanation of the Risk Drivers

Now we will present the risk drivers we'll be considering today. There will be a chance in a few minutes to discuss these risk drivers, so please hold comments and questions.

[*review slides—note that there may be new risk drivers based on the initial worksheets that we will need to explain*]

Discussion of COAs

I'd like to open the floor up for discussions about the definitions of the COAs we are considering today. Does anyone have any questions? Was there anything confusing or that needs to be added or deleted? At this point we are not discussing whether these COAs will be successful or not. We are only clarifying their definitions.

[allow for open discussion]

[if needed] From this discussion, it sounds as if we need to amend the definition for COA [Air Force/PBL/etc.] to say [new definition] *[note taker to note this for report]*

Discussion of Risk Drivers

I'd like to open the floor up for discussions about the definitions of the risk drivers we are considering today. Does anyone have any questions? Was there anything confusing or that needed to be added or deleted? At this point we are not discussing whether these risk drivers are relevant or important to the COAs. We are only clarifying their definitions.

[allow for open discussion]

[if needed] From this discussion, it sounds as if we need to amend the definition for risk driver [data/personnel/etc.] to say [new definition] *[note taker to note this for report]*

Let's also take a moment to discuss whether this list of risk drivers is comprehensive. That is, are we missing any risk drivers that are important for the success of the COAs?

[if needed] We did discuss a new risk driver earlier that was added to this list. But are there any others? *[if someone mentions a new one]* Why do you think this is important? Do others think this is also important?

Discussion of Methodology

[show slide] Today we'll be starting with COA [name] and talking through your assessment of that COA against the risk drivers. If you remember, you filled out a worksheet like the one shown on the screen.

For this worksheet, we define a *successful [implementation]* as one that would cause no harm to the fleet (i.e., result in equal availability to the existing structure). *Note that in this definition, a successful [implementation] may be one that increases the cost or worsens performance metrics other than availability—as long as availability remains the same as it is currently.*

The second column asks whether the risk driver is *relevant* to that COA. Here, you assess whether the risk driver has any influence on the likelihood of a successful [implementation]. If there is even the slightest possibility of an influence, it should be deemed as relevant.

The third column asks whether the *relevant* risk drivers are *partially substitutable*, or whether lacking a full level of that risk driver would limit the Air Force's ability to successfully [implement] because *nothing can substitute completely* for that risk driver. For instance, could a function be successfully [implemented] if only a partial set of relevant personnel were attracted and retained or if only a partial transfer of data occurred? If nothing can substitute for these risk drivers, they would be considered NOT partially substitutable. Or is the successful [implementation of a COA element] possible with only a partial set of relevant personnel because the lack of personnel can be made up for with the proper IT tools? If that is the case, then personnel would be partially substitutable

For the risk drivers that are partially substitutable, the fourth column asks you to rank the importance of each substitutable risk driver to successful [implementation]. A rank of one is the most important, two is the second-most important, and so on. If two risk drivers are equally important, you may rank them equally (i.e., as a tie).

For the likelihood columns, we define the term *sufficiency,* or the lowest level of a specific risk driver needed to ensure a successful [implementation] with equal availability.

For the relevant risk drivers, the fifth column asks for the likelihood of *never* being able to achieve *sufficiency* of the risk driver. For instance, how likely is it that the data will *never* be able to achieve sufficiency under a COA? When determining this likelihood, think about all of the scenarios that could lead to the data not being successfully [accessed by the government] and the likelihood of those scenarios. Remember, *successful* scenarios can include those that increase costs or reduce performance metrics other than availability.

For the relevant risk drivers, the sixth through eighth columns ask for the likelihood that the Air Force will achieve *sufficiency* of the risk driver for that COA in (a) less than two years, (b) between two and five years, or (c) more than five years. When determining these likelihoods, think about the least amount of time needed to achieve sufficiency of that risk driver (i.e., a sufficient level such that successful [implementation] will be highly likely). Also, think about the scenarios where it might take longer and how long that could take.

Remember that since the four likelihoods that you provide here represent every possible state of the world (i.e., never [implement], [implement] in less than two years, [implement] in two to five years, [implement] in more than five years), the likelihoods will add up to 100 percent, as shown in the ninth column.

Before we move on, are there any questions about how to fill out the worksheet?

COA Results

Now, I'll show you a summary of the results of the group for COA [name]. After I show you this, we'll go through this piece by piece and discuss them as a group. Please hold your comments until after I present all of the results.

Let's begin with **relevance**. This slide shows, for each risk driver, the number of you who said the risk driver was and was not relevant. [*review specific results*]

Next, this slide shows, for each risk driver, the number of you who said the risk driver was and was not **partially substitutable**. [*review specific results*]

Next, this slide shows, for each risk driver, the average **rankings** you provided and the minimum and maximum value for the group. [*review specific results*]

Now, let's look at the **likelihood** for each risk driver. This slide shows a summary of the likelihood for each. For each likelihood, we show the average for the group, as well as the minimum and maximum value for the group. Note that because we are showing averages, the likelihood will not add up to 100 percent. [*review specific results*]

Now, let's combine some of this information. If we take into account all of the information you provided, we can get likelihoods for COA [name] for when ***initial and full implementation*** may take place. For initial implementation, we are referring to the time it takes to get at least one of the risk drivers in place, or in the case that some of the risk drivers are not substitutable, the time to get all nonsubstitutable risk drivers in place. For full implementation, we are referring to the time it takes to get all of the risk drivers in place. *[review specific results]*

Discussion of COA Results

Now we'll have a discussion about your answers. Because time is limited, we are going to start with the risk drivers that a majority of you rated as not substitutable, and then we'll move on to those that you ranked, on average, as very important. We'll go until we run out of time, saving a few minutes to look over the results.

Relevance

Let's start with risk driver X. If we go back to the **relevance** slide, I see that [everyone agreed/there was disagreement] about this being relevant.

[if disagreement] Could someone please tell me why they thought this risk driver was relevant? Could someone please tell me why they thought it wasn't relevant? After hearing both sides of the argument, does anyone think they may have changed their mind?

Substitutable

Let's now look at the substitutable slide. I see that [everyone agreed/there was disagreement] about this being partially substitutable.

[if disagreement] Could someone please tell me why they thought this risk driver was partially substitutable? Could someone please tell me why they thought it wasn't partially substitutable? Is there anything else that someone thinks could at least partially substitute for this risk driver?

[if group unanimously thinks it is not partially substitutable] Can anyone think of anything that would be at least a partial substitute for this risk driver?

Ranking of Importance

Let's now look at the ranking slide.

[if determined to not be partially substitutable] Since we just determined that this was not partially substitutable in the last section, we don't have to rank this one. However, let's talk briefly about how important it is relative to the other risk drivers. Could someone please tell me why they think this risk driver is really important or more important than other risk drivers? Could someone please tell me why they think it isn't important or less important than other risk drivers?

[if determined to be partially substitutable] I see that [everyone agreed/there was disagreement] about how important this risk driver is.

[if disagreement] Could someone please tell me why they thought this risk driver was really important or more important than other risk drivers? Could someone please tell me why they thought it wasn't important or less important than other risk drivers?

[if group unanimously agrees on importance] Can anyone think of any reason why or any situation where one of the lower-ranked risk drivers may be more important than this one? Where one of the higher-ranked risk drivers may be less important than this one?

Likelihoods

Let's now look at the likelihoods. I see that everyone's answers varied about the likelihood that this risk driver will never reach sufficiency.

Most people thought [it should take [<2, 2–5, >5]; it would never] for the risk driver to reach sufficiency. Can someone who thought that explain why to the group? Can someone who thought it would take [less/more] time explain why they thought that? I see [no one/very few people] thought that it was most likely to take [<2, 2–5, >5]. Can anyone think of a scenario where it would actually take this long?

[if disagreement] The average likelihood here was X. Could someone please tell me why they thought the likelihood of this risk driver never being able to reach sufficiency was less than X? More than X? Can someone tell me about a scenario in which this risk driver would never reach sufficiency?

Revise Risk Worksheets

Taking into account everything you learned over this session, I would like you to take the last few minutes of this workshop to revisit your initial answers *[hand out initial worksheets and new blank worksheets]*. You may decide to answer the same way or to change your answers. Please fill out the new worksheet and then hand it in to me.

Thanks again for your support of the F-22 BCA Risk Assessment. We all appreciate your time and effort! The results of these risk assessments will be presented in the BCA final report in September.

Acknowledgments

Numerous people both within and outside the Air Force provided valuable assistance to and support of our work. They are listed here with their ranks and positions as of the time of this research. We thank Maj Gen Dwyer Dennis and Brig Gen Eric Fick, former and current program executive officers for Fighters and Bombers, Air Force Life Cycle Management Center, for sponsoring this work. We also thank Col Anthony Genatempo, Air Force Life Cycle Management Center, Fighters & Bombers Directorate, F-22 Division, and Lt Col Rodney Stevens, Air Force Life Cycle Management Center, Fighters & Bombers Directorate, F-22A Division, PM-Sustainment Branch, for their tireless support throughout this research. We also thank their staffs for their time and support during this research.

We wrote this document as part of a much larger effort to support the F-22 System Program Office's 2015 product support BCA for the F-22. We benefited from repeated discussions with RAND colleagues involved in that effort. Michael Boito and Kristin Lynch led that effort under the close and constructive oversight of program and deputy program directors Obaid Younossi and James Powers. Other RAND colleagues on the team included John Drew, Guy Weichenberg, and Emma Westerman. Meg Harrell, Laura Miller, and Laura Werber provided valuable advice on how to collect and document professional judgments relevant to our risk analysis. Albert Robbert provided valuable insights on how to estimate indirect Air Force costs. RAND colleagues Sean Bednarz, Natalie Crawford, Eric Peltz, Lara Schmidt, Don Snyder, and, before he became program director, Obaid Younossi served constructively on an in-house RAND quality assurance panel for our BCA support project. Consultants Larry Klapper and John Spicer provided detailed information on methods used and approved in past DoD BCAs and on performance-based logistics agreements. Communication analysts Barbara Bicksler and Jerry Sollinger provided close editorial support throughout the BCA support effort. Megan Bishop, Darlette Gayle, Michelle Horner, and Tandrea Parrott provided able administrative support.

Our formal efforts to elicit professional judgments from subject-matter experts in many organizations helped us refine this methodology. We also briefed a variety of audiences on this methodology and benefitted from their questions and recommendations. These personnel came from Air Force Materiel Command Logistics, Engineering, and Force Protection; Strategic Plans, Programs, Requirements, and Assessments; and Financial Management. They also came from the Air Force Sustainment Center Logistics Directorate, Cost Division, and 448 Supply Chain Management Wing; the Air Force Life Cycle Management Center Logistics Directorate; the Secretary of the Air Force Logistics and Product Support, Global Power Leadership, and Financial Management offices; Air Combat Command, Directorate of Maintenance and Logistics and Directorate of Plans, Programs, and Requirements; Pacific Air Forces/Logistics; Air Force

Logistics, Engineering, and Force Protection; the Defense Logistics Agency; the Air Force Cost Analysis Agency; Lockheed Martin Aeronautics; Boeing; and Pratt & Whitney.

We thank all who have advised and supported us as we have applied and refined this methodology. We retain full responsibility for any parts of it that can still be improved.

Abbreviations

AIP	Aircraft Integrity Program
ASM	Aircraft Sustainability Model
BCA	business case analysis
COA	course of action
CY	calendar year
DoD	U.S. Department of Defense
FY	fiscal year
IT	information technology
LO	low observable
NPV	net present value
OMB	Office of Management and Budget
PAF	Project AIR FORCE
PBL	performance-based logistics
SME	subject-matter expert

References

Adler, Matthew D., and Eric A. Posner, eds., *Cost-Benefit Analysis: Economic, Philosophical, and Legal Perspectives*, Chicago, Ill.: University of Chicago Press, 2001.

Arrow, K. J., H. B. Chenery, B. S. Minhas, and R. M. Solow, "Capital-Labor Substitution and Economic Efficiency," *Review of Economics and Statistics,* Vol. 43, No. 3, August 1961, pp. 225–250.

ASM® Sparing Model, McLean, Va.: Logistics Management Institute, 2012.

Ata, Mustafa Y., "A Convergence Criterion for the Monte Carlo Estimates," *Simulation Modelling Practice and Theory*, Vol. 15, No. 3, March 2007, pp. 237–246.

Campbell, Harry F., and Richard P. C. Brown, *Benefit-Cost Analysis: Financial and Economic Appraisal Using Spreadsheets*, Cambridge, UK: Cambridge University Press, 2003.

Clemen, Robert T., and Terence Reilly, *Making Hard Decisions*, Cengage Learning, 3rd edition, 2014.

Cox, Louis Anthony, Jr., "What's Wrong with Risk Matrices?" *Risk Analysis*, Vol. 28, No. 2, April 2008, pp. 497–512.

Cox, Louis Anthony, Jr., Djangir Babayev, and William Huber, "Some Limitations of Qualitative Risk Rating Systems," *Risk Analysis*, Vol. 25, No. 3, June 2005, pp. 651–662.

Dawes, Robyn M., *Rational Choice in an Uncertain World*, New York: Harcourt Brace Jovanovich, 1988.

Director, Cost Assessment and Program Evaluation, "Economic Analysis for Decision-Making," DoD Instruction 7041.03, Washington, D.C.: Department of Defense, September 9, 2015.

Fishman, George S., *Monte Carlo: Concepts, Algorithms, and Applications*, New York: Springer-Verlag, 1996.

Hertz, David B., "Risk Analysis in Capital Investment," *Harvard Business Review*, Vol. 42, No. 1, February 1964.

Holtan, Marius, "Using Simulation to Calculate the NPV of a Project," InvestmentScience.com, May 31, 2002. As of August 9, 2016:
http://www.investmentscience.com/Content/howtoArticles/simulation.pdf

Keeney, Ralph L., *Value-Focused Thinking*, Cambridge, Mass.: Harvard University Press, 1996.

Law, Averill M., and W. David Kelton, *Simulation Modeling and Analysis*, 3rd ed., New York: McGraw Hill, 2000.

MacKenzie, Cameron A., "Summarizing Risk Using Risk Measures and Indices," *Risk Analysis*, Vol. 34, No. 12, 2014, pp. 2143–2162.

McFadden, Daniel, "Constant Elasticity of Substitution Production Functions," *Review of Economic Studies*, Vol. 30, No. 2, June 1963, pp. 73–83.

Morgan, M. Granger, Baruch Fischoff, Ann Bostrom, and Cynthia J. Atman, *Risk Communication: A Mental Models Approach*, Cambridge, UK: Cambridge University Press, 2002, reprinted 2011.

Office of Management and Budget, *Guidelines and Discount Rates for Benefit-Cost Analysis of Federal Programs*, Circular No. A-94 (revised), Washington, D.C., 1992.

———, "Discount Rates for Cost-Effectiveness, Lease Purchase, and Related Analyses, Appendix C (revised)," December 2014, in *Guidelines and Discount Rates for Benefit-Cost Analysis of Federal Programs*, Circular A-94 (revised), Washington, D.C., 1992. As of August 11, 2015:
https://www.whitehouse.gov/omb/circulars_a094/a94_appx-c

Office of the Principal Deputy Assistant Secretary of Defense for Logistics and Materiel Readiness, *DoD Product Support Business Case Analysis Guidebook*, Washington, D.C.: Office of the Secretary of Defense, 2011.

OMB—*see* Office of Management and Budget.

Rhea, Bob, Steven Hurt, Alan Heckler, Sameer Dohadwala, Hamza Rampurawala, and Rajan Singh, *Recommendation for Long-Term Sustainment of the F-22 Raptor, F-22 Sustainment Business Case Analysis*, Chicago, Ill.: A. T. Kearney, October 2009.

Saaty, Thomas L., *The Analytic Hierarchy Process: Planning, Priority Setting, Resource Allocation*, New York: McGraw Hill, 1980.

Savvides, Savvakis, "Risk Analysis in Investment Approach," *Project Appraisal*, Vol. 9, No. 1, March 1994, pp. 3–18.

Sherbrooke, Craig C., *Optimal Inventory Modeling of Systems: Multi-Echelon Techniques*, New York: John Wiley and Sons, 1992.

Slay, F. Michael, and Randall M. King, *Prototype Aircraft Sustainability Model*, Report AF601R2, McLean, Va.: Logistics Management Institute, 1987.

Uzawa, Hirofumi, "Production Functions with Constant Elasticities of Substitution," *Review of Economic Studies*, Vol. 29, No. 4, October 1962, pp. 291–299.

Wallsten, Thomas S., and David V. Budescu, "State of the Art—Encoding Subjective Probabilities: A Psychological and Psychometric Review," *Management Science*, Vol. 29, No. 2, February 1983, pp. 151–173.